# Baofeng Radios - A Bootcamp for Beginners

Unbox to First Contact in 60 Minutes, Menus Made Simple, Repeater + Channel Programming, First-Call Script, and Troubleshooting Trees for Emergency Comms

MorseCode Publishing

For permissions, bulk sales, or special inquiries, contact:
Jared@MorseCodePublishing.com

# Contents

# Disclaimer: Safety, Legality, and Common Sense

Let's get the serious stuff out of the way so we can get to work.

The Baofeng UV-5R (and its many clones) is an incredible tool, but it is not a walkie-talkie from a toy store. It is a powerful radio capable of transmitting on frequencies reserved for amateur radio operators, emergency services, and commercial businesses.

**Here is the golden rule:** You can *listen* to almost anything legally. But you cannot *transmit* (press the push-to-talk button) on amateur radio bands without an amateur radio (Ham) license issued by your local government (like the FCC in the United States).

This book is a practical training guide, not legal advice.

- Do not use this book as an excuse to transmit illegally.
- We will not provide "secret frequency lists" for rule-breaking.
- We highly encourage you to get your basic amateur radio license, it's easier than you think, and it unlocks the full power of your radio.

For anything related to local laws, frequencies, or emergency use exemptions, consult official sources and your local radio clubs. Now, let's learn how to use this thing.

# Structure of a UV-5R Radio

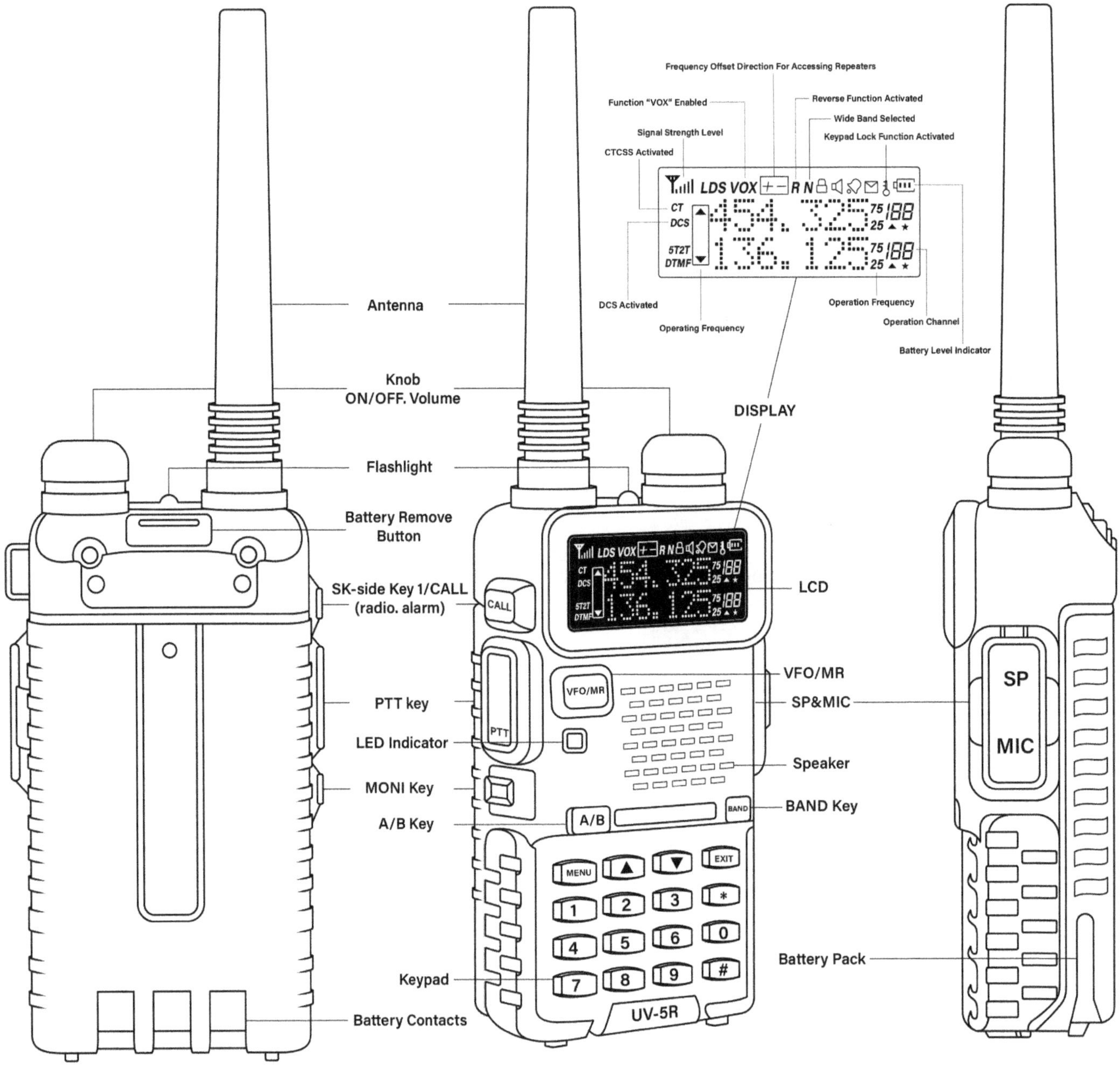

# How to Use This Book

You bought this radio for peace of mind, preparedness, or as your first step into a new hobby. But right now, you're staring at a little screen with confusing numbers, and the manual that came in the box reads like a poorly translated physics textbook.

Take a breath. **You can do this.** You don't need to be an engineer to use a Baofeng. You just need a good coach. We have designed this book to deliver on one specific promise: *You can go from unopened box to hearing traffic, saving channels, and making a clean first transmission (legally) in exactly one hour.*

To make this easy, this book is split into three specific paths. You are in control of the journey.

## Path 1: The First Hour (Start Here)

If you just took the radio out of the box, start at Chapter 1. We will ignore 90% of the useless menus and focus only on the buttons that get you hearing voices and making your first transmission within 60 minutes.

## Path 2: Fix It Fast

Already messed up your settings? Radio making a weird noise? Can't hear anything? Jump to **Part 2: Troubleshooting**. It is a symptom-based triage tree. You look up your problem, and we give you the 60-second fix. No theory, just solutions.

## Path 3: The Family Plan

If you bought a multi-pack of radios to keep your family connected during a blackout, cell outage, or camping trip, jump to **Part 3**. We'll show you how to set up a stress-proof communication plan that anyone in your family can use, even if they've never read this book.

## Gear Checklist

You do not need a lot of expensive gear to be successful. Let's look at what you actually need right now.

**The Minimum (Start With This):**

- Your Baofeng UV-5R radio.
- A fully charged battery.
- The stock "rubber duck" antenna that came in the box.

**Strongly Recommended:**

- **A spare battery and charger.** Radios always seem to die right when you figure them out. Having a backup keeps the momentum going.

**Optional (Nice to Have Later):**

- **A programming cable and CHIRP software.** This lets you program the radio from your computer. It's amazing, but we won't need it for your first hour. We are going to learn how to drive the radio manually first.

**[FIELD NOTE] Avoid wasting money on...** Giant, 15-inch "tactical" whip antennas. Yes, they look cool, and yes, they give you a bit more range. But right now, they just make the radio top-heavy and clumsy to hold while you are trying to learn the buttons. Stick to the stock antenna for your first week.

## Your Quick-Start "Mission Map"

Keep this page bookmarked. This is your itinerary for the next 60 minutes. We are going to build your skills layer by layer.

**Minute 0–5: Power, Volume, and Silence** We'll turn it on, set a comfortable volume, and adjust the "squelch" so the radio stops making that horrible static hissing noise.

**Minute 5–15: Find a Lighthouse** Before we try to talk to anyone, we need to prove the radio works. We will find a local, reliable signal (like a weather broadcast) to listen to.

**Minute 15–35: The Repeater Puzzle** We'll demystify "repeaters", the towers that give your little radio massive range. We'll translate confusing terms like *offsets* and *tones* into plain English.

**Minute 35–55: Program and Save** This is where most people quit. You won't. We will manually program a repeater into your radio and save it to a memory channel so you never have to type it in again.

**Minute 55–60: Your First Transmission** You'll press the button. We'll give you a simple, embarrassment-proof script to use for your very first legal transmission.

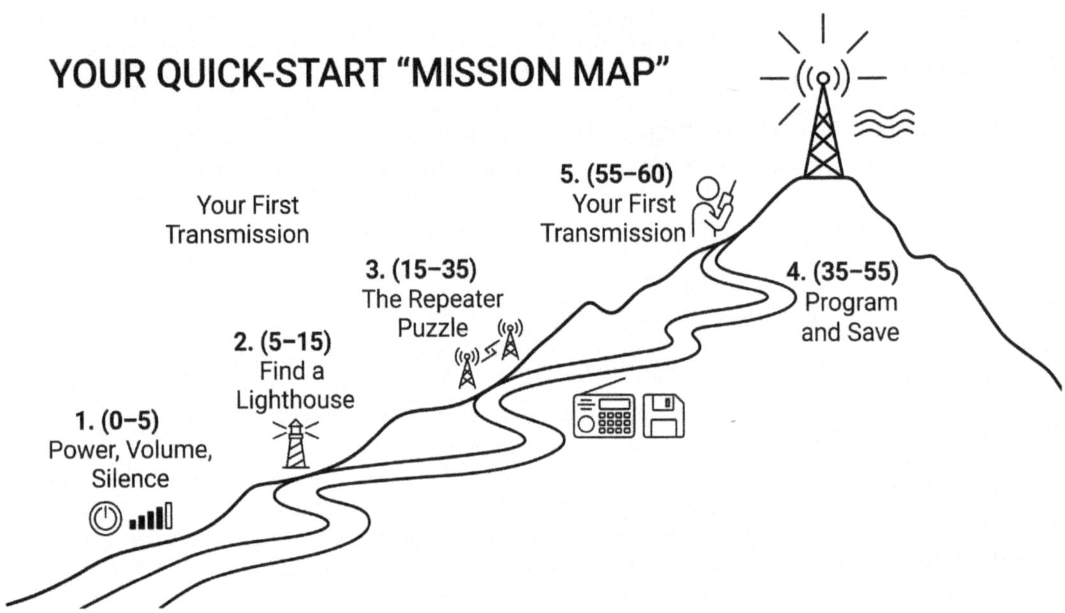


Ready? Turn the page, and let's start the clock on your First Hour.

- Jared Johnson (KF0RTU) – Jared@MorseCodePublishing.com

# Part 1:

## The First Hour (Get Working Fast)

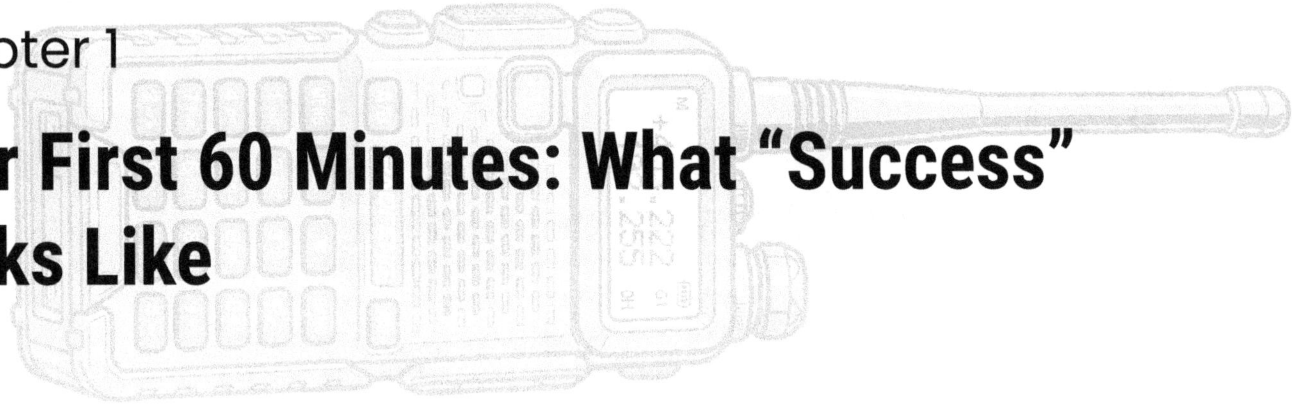

Chapter 1

# Your First 60 Minutes: What "Success" Looks Like

You bought the radio. Now it's just beeping at you. The screen is flashing randomly, a robotic voice prompt is yelling numbers every time you brush a key, and you're afraid to press a button in case you accidentally call in a rescue helicopter or break a federal law.

Relax. Take your hand off the radio, take a deep breath, and let's get oriented. You are exactly where every beginner starts.

**What you'll accomplish:**

- Define the 3 wins so you know exactly what we are trying to achieve today (and what we can safely ignore).
- Understand VFO vs. Memory (this is the single most important concept in this entire book).
- Prevent the 5 most common beginner mistakes that cause people to throw their radios in a drawer and quit.

**What you need:**

- Your Baofeng UV-5R (keep it powered off for now so it doesn't distract you).
- A comfortable place to sit.

## The 3 Wins (What Success Feels Like)

When you look at a radio with 21 buttons and a screen full of tiny icons, it's incredibly easy to get overwhelmed. We are going to ignore almost all of them today. For your first 60 minutes, we only care about achieving three specific, measurable wins:

1. **Hear:** We need to prove the radio works out of the box by catching a strong, local signal. Hearing traffic validates your equipment and builds your confidence before you ever try to transmit.
2. **Store:** We are going to find a local frequency and save it to the radio's permanent memory. You are going to build a "rolodex" of safe channels so you never have to blindly type numbers in an emergency.
3. **Talk:** You are going to make a short, clean, legal transmission (using your ham radio license) without freezing up or feeling stage fright.

That's it. If you can hear, store, and talk, you have mastered 90% of what this radio was built to do. Everything else in the manual is just optional bells and whistles.

## The Two Modes (VFO vs. Memory)

If you only remember one thing from this chapter, let it be this: **Your radio has a split personality.** It exists in one of two distinct modes, and understanding the difference is the secret key to unlocking the Baofeng.

FIG 2

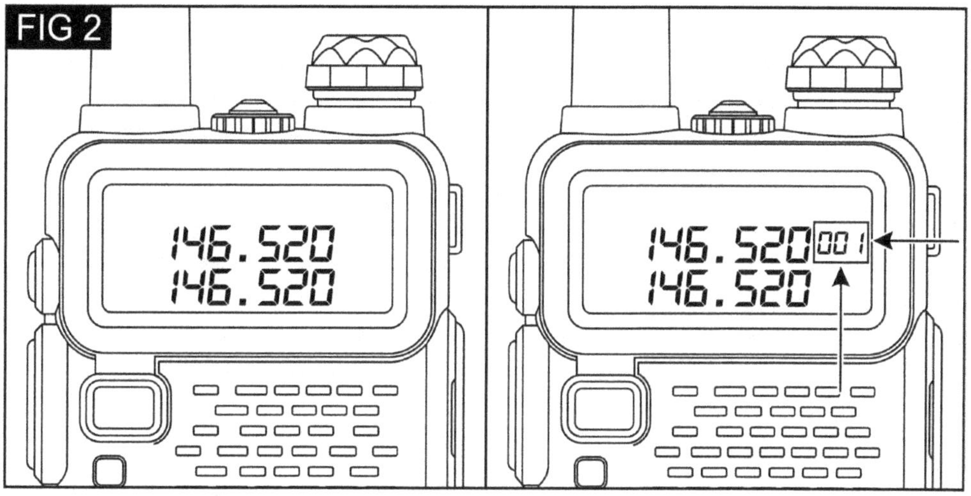


VFO MODE (Frequency) MEMORY MODE (Channel)

**Mode 1: VFO Mode (Frequency Mode)** Think of VFO mode like the manual tuning dial on a classic car radio. You are turning the knob, searching through the static to see what is out there. In this mode, you use the keypad to type in exact, raw frequencies (like 1 6 2 . 5 5 0 for a local National Weather Service broadcast). You are freely exploring the airwaves. *Crucially, you must be in VFO mode to program new channels into the radio.*

**Mode 2: Memory Mode (Channel Mode)** Think of Memory mode like the preset push-buttons on your car radio. Once you find a frequency you like in VFO mode, you save it as a "Channel" (like Channel 1, Channel 2, etc.). When you are in Memory mode, the danger of accidentally typing the wrong number disappears. You just flip safely through your pre-saved, trusted channels.

**[FIELD NOTE] How to switch between them:** On the UV-5R, there is an orange button labeled **VFO/MR**. Pressing this button toggles the radio back and forth between manual tuning (VFO) and your saved presets (Memory/MR).

## Why You Feel Lost (The Baofeng Reality)

If you feel confused by the menu system, it's not your fault. These radios were originally designed by engineers, for other engineers. The menus are not alphabetical, the abbreviations on the screen don't always make obvious sense, and the factory manual is famously unhelpful. The radio provides zero context clues to tell you what a button actually does.

Don't let this intimidate you. Put the factory manual in a drawer right now, you don't need to memorize the menus to be highly effective. In this book, we will give you the exact, step-by-step keypad combos you need. Treat the radio like a calculator: you don't need to know how the circuit board works. Just punch in the exact formula we give you, and you will get the right result every single time.

## The 5 Time-Wasters

Thousands of beginners have walked this exact path before you. We know exactly where they stumble. Keep an eye out for these 5 common traps, we will solve all of them together in the coming chapters:

1. **Tone Confusion:** If you grew up using cheap walkie-talkies from a sporting goods store, you probably learned about

"privacy codes." On a Baofeng, trying to use these tones (CTCSS/DCS) when you don't actually need them results in nobody being able to hear you.

2. **Wrong Mode:** The frustration of typing a raw frequency while stuck in Memory mode and watching the radio ignore you, or trying to save a channel while in Memory mode.
3. **Wrong Step Size:** Typing in a frequency and watching the radio magically change the last digit on you, feeling like the machine is fighting your inputs.
4. **Saving Order:** The Baofeng requires you to save the listening frequency and the talking frequency in a very specific, quirky order. It isn't intuitive. If you mess up the order, the channel simply won't work.
5. **Squelch Too High:** The "squelch" is the radio's static filter. Setting this invisible gate so high that the radio goes completely deaf is the number one reason beginners think their new radio is broken.

**[STOP] Don't push buttons just to see what happens.** If you dive into the menu and start changing random settings to see what they do, you will eventually lock yourself out of transmitting or receiving. Stick to the guide.

## Quick Orientation: The Big Four

Before we actually turn the radio on in the next chapter, physically locate these four things on the face of your device:

- **The Keypad:** This feels like an old flip phone. It is used for typing frequencies directly and selecting numbered menu options.
- **The MENU Button:** Located in the center of the up/down arrows. This acts as your "Enter," "Select," or "Confirm" key.
- **The A/B Display:** Your screen has a top line (A) and a bottom line (B). The radio can monitor two frequencies at once, but you can only actively control one line at a time. The blue **A/B button** flips your control between the top and bottom. You will know which line is active by the small black triangle pointing to it on the left side of the screen.
- **The PTT Button:** This is the large, tactile rubber button on the left side of the radio housing. PTT stands for "Push To Talk." *Do not press this today until we explicitly tell you to.*

## Quick Win Checklist

- [ ] I understand the critical difference between VFO mode (manual typing) and Memory mode (saved presets).
- [ ] I know that I use the orange VFO/MR button to switch back and forth between the two modes.
- [ ] I have located the MENU button (which acts like the "Enter" key).
- [ ] I have located the A/B button (which switches my control between the top and bottom lines of the screen).
- [ ] I have located the big rubber PTT button on the side, but I will strictly avoid pressing it for now.

## If you're stuck, go to...


- *Chapter 10: I Can't Hear Anything* (if you are playing with the radio early and it's dead silent).
- *Chapter 12: My Channels Won't Save* (if you forgot the rule about VFO vs. Memory).

Chapter 2

# Five-Minute Setup

## Power, Volume, Squelch, and a Clean Receive

YOU CAN'T FIX WHAT you can't hear. If your radio is completely silent, or if it is constantly blasting you with a wall of ear-piercing static, you are never going to catch a signal. Before we can even think about programming channels, we need to get the physical hardware running smoothly so you have a clean, reliable baseline.

**What you'll accomplish:**

- Set your volume and "squelch" correctly (finding the perfect Goldilocks zone for listening).
- Use the physical Monitor button to prove your speaker actually works.
- Learn how to lock the keypad so you don't accidentally ruin your progress.

**What you need:**

- Your Baofeng radio.
- The stock antenna (screwed firmly into the top).
- A charged battery.

### Step 1: The Hardware Check

Before we turn the knob, we need to assemble the hardware correctly.

1. **Seat the battery:** Slide the battery onto the back of the radio, pressing upward until you hear a firm, satisfying *click*. Don't be afraid to use a little force with your thumbs, brand new radios often have very stiff plastic rails. If the battery is loose, the radio will randomly shut off right in the middle of a transmission.
2. **Attach the antenna:** Screw the stock rubber antenna onto the top brass connector. Thread it until it is snug, but do not crank it down with superhero strength.

**[STOP] Never press the Push-To-Talk (PTT) button if the antenna is unscrewed.** Transmitting without an antenna creates a backlog of power that has nowhere to go. It will reflect right back into the radio's circuit board and literally fry the internal components. Always have an antenna attached.

### Step 2: Power and Volume

Turn the volume knob on the top of the radio clockwise. You will feel a mechanical *click* as the radio powers on.

The screen will light up, flashing an orange or purple backlight, and a loud robotic voice will likely announce "FREQUENCY MODE" or "CHANNEL MODE."

Turn the knob about halfway up. Do not crank it to maximum volume just yet, when a signal does come through, the Baofeng speaker is surprisingly loud and can easily startle you. Right now, you probably won't hear anything else, which leads us to the most important testing button on the side of the radio.

### Step 3: The Monitor Button (Proving the Speaker Works)

Right below the big PTT button on the left side of the radio, there is a smaller button labeled **MONI** (Monitor).

Press and hold the **MONI** button right now.

You should immediately hear a loud, rushing hiss of white noise. This is the raw sound of empty radio waves. While holding the button, adjust your top volume knob until the static is at a comfortable listening level.

Let go of the MONI button. The static should instantly stop. *Congratulations, you just proved your speaker and receive path are working perfectly.*

### Step 4: Setting the Squelch (The Invisible Gate)

Why did the static stop when you let go of the MONI button? Because of the **Squelch**.

The squelch is like an invisible dam that holds back the constant river of background static in the atmosphere. When the squelch is working correctly, your radio stays dead silent. But when a real, powerful signal (like someone talking) hits your antenna, it "breaks the squelch", spilling over the dam and opening the speaker so you can hear them.

If your squelch is set too low (Level 0), the dam is gone, and you hear non-stop hissing. If your squelch is set too high (Level 9), the dam is too tall. Even strong, local signals won't be able to break through, and your radio will seem totally deaf.

Let's set your squelch to the perfect beginner level (Level 1 or 2):

1. Press the **MENU** button.
2. Type the numbers **0** then **0** on the keypad. Your radio has 40 different numbered menus, and Menu 0 is always Squelch. The screen should now say SQL (Squelch).
3. Press the **MENU** button again. You will see a small arrow drop down to the bottom line of the screen. You are now unlocked and editing the setting.
4. Type the number **2**.
5. Press the **MENU** button one more time to save it. You will hear the robotic voice say "Confirm."
6. Press the **EXIT** button to leave the menu.

You just navigated your first Baofeng menu! You used the exact "formula" you will use for the rest of this book: Menu, find the setting, Menu to unlock, change it, Menu to save, Exit.

## Step 5: Lock it Down

Once you have your radio set up exactly the way you like it, it is incredibly easy to accidentally bump a keypad button with the palm of your hand, or have it rub against the inside of your jacket pocket, completely changing your frequency without you noticing.

Look at the bottom right key on your keypad. It has a # symbol and a tiny picture of a padlock. Press and hold the **# / Padlock key** for about two seconds. You will hear a beep, and a tiny padlock icon will appear at the top center of your screen. Your keypad is now locked. You can still use the volume knob, and you can still press the PTT button to talk, but you cannot accidentally change channels or dive into menus. To unlock it, just press and hold the # key for two seconds again until the padlock icon disappears.

*[GOOD ENOUGH]* If your battery is seated, your antenna is on, your squelch is set to 2, and you've locked your keypad so you don't accidentally ruin things, you are in perfect shape. Your radio is primed and ready to catch its first real signal.

## Quick Win Checklist

- [ ] My antenna is firmly attached to the radio.
- [ ] I understand that I should never press the PTT button if the antenna is removed.
- [ ] I have used the side MONI button to hear raw static and adjust my volume knob to a comfortable level.
- [ ] I understand that "Squelch" acts as an invisible gate to block background static.
- [ ] I have successfully navigated the menu and set my Squelch (SQL) to Level 2.
- [ ] I know how to lock and unlock my keypad using the # key.

## If you're stuck, go to...

- *Chapter 10: I Can't Hear Anything* (if you pressed the MONI button and heard absolutely zero sound).

# Narrative Interlude: The Beeping Brick

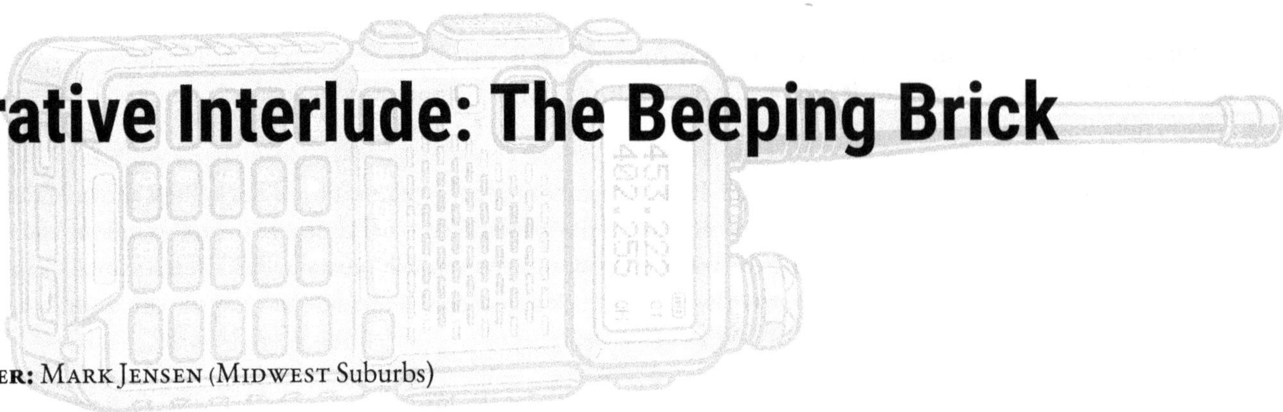


**Character:** Mark Jensen (Midwest Suburbs)

**Skills reinforced:**

- VFO vs. Memory Mode
- Volume and Squelch setup
- Keypad Lock
- Setting "Beginner-Friendly" Defaults

**Where to learn the steps:** Chapters 1 & 2

The box sat on the center of the kitchen island like a tactical alien artifact. Mark Jensen, age forty-two and officially a licensed Amateur Radio Technician as of three days ago, stared at it with a mixture of pride and growing dread. To the uninitiated, it was a thirty-dollar piece of electronics; to Mark, it was a gateway to the global airwaves, a piece of gear that signaled he was a man with a plan for the next blizzard.

"Is it supposed to look that... small?" his wife, Jen, asked, leaning over a bowl of salad. She poked the stubby antenna through the plastic wrapping.

"It's compact, Jen. Ruggedized," Mark corrected, though he was secretly thinking the same thing. In the online photos, it looked like something a paratrooper would carry; in his kitchen, next to a bowl of cherry tomatoes, it looked like a high-tech garage door opener. "It's a masterpiece of Chinese engineering. Industrial. Utilitarian. It's a Baofeng UV-5R."

"It looks like a walkie-talkie I had in 1994," his son, Owen, chimed in, finally looking up from his phone with a glimmer of teenage skepticism. "Does it have Snake on it? Or at least a flashlight?"

"It has the power to communicate across the Twin Cities during a grid-down scenario, Owen," Mark said, trying for a tone of steady, military-grade competence. "And yes, it actually does have a flashlight." He reached for the box, carefully extracting the radio from its crinkly plastic cocoon.

The unboxing started well. The battery clicked onto the rails with a satisfying mechanical snap that sounded like a magazine being loaded into a rifle. The antenna screwed on smoothly. Mark felt his confidence rising, until he twisted the top knob.

*"BEEP. FREQUENCY MODE."*

The radio didn't just talk; it shouted. A loud, female, digitized voice echoed off the granite countertops, sounding like a very stern flight attendant.

"Whoa!" Owen jumped. "Is that the emergency? Is the world ending right now?"

"No," Mark muttered, his face reddening. He tried to turn the volume down, but his fingers were slightly sweaty, and he accidentally grazed the keypad.

*"BEEP. SIX. BEEP. SEVEN. BEEP. MENU."*

The radio was now screaming random integers at the ceiling. Mark frantically pushed a button, he thought it was the 'Exit' button, but instead, the radio emitted a long, sustained, screeching siren tone that sounded like a smoke alarm having a nervous breakdown in the middle of a library.

"Mark, honey," Jen said, her voice strained as she covered her ears. "Make it stop. Please."

Mark mashed the orange button. Then the blue one. Then he just started clicking the top knob back and forth like he was trying to win a game of arcade whack-a-mole. Finally, he managed to twist the power off. The silence that followed was heavy and judgmental.

"Okay," Mark said, clearing his throat and adjusting his glasses. "That was just... a diagnostic check. Calibrating the audio levels. It's an advanced feature."

"You looked like you were trying to disarm a bomb and forgot which wire to cut," Owen said, returning his gaze to his phone. "I give it ten minutes before the FBI knocks on the door to see why you're broadcasting a dial-up modem sound to the neighbors."

Mark ignored him and turned the radio back on, moving with the agonizing caution of a man handling nitroglycerin. He kept the volume at a whisper. But now, as he stared at the dual-line display, there was a new problem. The radio was dead silent. He could see two lines of numbers, 144.000 on top and 440.000 on the bottom, but even when he turned the volume knob all the way to the right, nothing happened. No static, no voices, just a cold, digital stare.

"I think it's broken," Mark whispered. "It was beeping a second ago, and now it's just... dead. I must have fried the circuit board when I was button-mashing."

He felt that familiar beginner's spiral. He'd had the radio for exactly four minutes and he'd already turned it into a tactical paperweight. He tried typing in the National Weather Service frequency he'd seen in the book, 162.550, but every time he hit the '1', the radio just let out a sharp, dismissive beep and stayed stubbornly on 144.000.

"It won't even let me type," he groaned. "It's locked. Or haunted. I've bought a haunted radio."

Jen walked over, wiping her hands on her apron. She picked up the field manual Mark had left on the counter, the one with the bright yellow 'Mission Map' on the front.

"Did you actually do the five-minute setup?" she asked, pointing to a section labeled 'The Invisible Gate.'

"Jen, I'm a licensed ham. I have a callsign. I studied for weeks. I don't need to, "

"Mark," she interrupted gently, tapping a finger on the page. "Read Step 2."

Mark sighed and looked at the page. *The Squelch Wall.* He looked back at his radio. He realized that in his earlier panic, he'd dived into the menu and changed 'SQL' to 9, thinking higher numbers meant more 'power' or better 'quality.' Instead, he'd built a digital wall so thick that not even a lightning bolt from the gods could break through to the speaker.

"Wait," Mark said, his curiosity finally overriding his embarrassment. He followed the instructions. *Menu -> 0 -> Menu.* He dropped the squelch from 9 down to a sensible 2. *Menu -> Exit.*

Nothing. Still silence.

"It's still dead, Jen. The wall is down, but nobody's home."

"Try the side button," she suggested, pointing at the checklist again. "The Monitor one. It says it bypasses the gate entirely."

Mark pressed the small, rubberized button labeled **MONI** located just below the large Push-To-Talk button.

*SHHHHHHHHHHHHHHHHH!*

A glorious, rushing waterfall of white noise exploded from the speaker. He jumped, then laughed, a huge grin breaking across his face. "Okay! The speaker isn't fried. It's alive! That's the sound of the universe right there."

"Now look at the right side of the screen," Jen said, reading ahead. "Does it have tiny numbers next to the frequencies?"

Mark squinted at the display. "Yeah. A little 001 and 002. What does that mean?"

"The book says that's Memory Mode. You're in the 'locked display case.' You can only use what's already saved there. To type new numbers, you have to push the orange button to get to the 'VFO Canvas.'"

Mark pressed the orange **VFO/MR** button. The tiny 001 vanished. The radio felt different, unlocked. He tentatively typed the digits: 1 6 2 5 5 0.

*"...mostly cloudy with a sixty percent chance of rain tonight. Lows in the mid-forties, winds from the North-Northwest at ten miles per hour..."*

The robotic, rhythmic voice of the National Weather Service filled the kitchen. It was clear, steady, and incredibly satisfying. Mark stood a little taller, adjusted his posture, and looked at his son.

"Look at me," he said, gesturing to the radio. "I'm basically NASA. Mission Control is officially online."

"You're a guy listening to a weather report in his kitchen, Dad," Owen said, though he actually looked a bit impressed that the beeping had been replaced by actual information. "But hey, at least it's not screaming at us anymore. That's a win."

Mark realized he'd spent the last ten minutes fighting the radio because he was trying to run before he could walk. He'd assumed silence meant hardware failure, when it really just meant he'd closed the gate too tight and was standing in the wrong room.

He held down the # key for two seconds. A tiny padlock icon appeared on the screen with a cheerful *chirp*.

"There," Mark said, setting the radio down. "Keypad lock. Now nobody, not even me, can accidentally mess up the settings while I'm making coffee."

"I'm not the one who turned it into a smoke alarm," Owen reminded him.

Mark just smiled, listening to the weather bot. He had his first win. He'd navigated the menus, conquered the squelch, and proven the receiver worked. The beeping brick was now a functioning tool.

"Okay," Mark said, looking at the next chapter in the guide. "I'm hearing the lighthouse. Now I just need to figure out where the other ships are."

## Takeaway

Silence usually isn't a broken radio, it's just a setting out of place. Most beginner frustration comes from being in the wrong Mode (Memory vs. VFO) or having the Squelch gate set too high. Use the Monitor button to prove your hardware is okay, then check your mode.

**Bridge:** Mark's radio is receiving loud and clear. But to talk to someone miles away, he needs to understand the towers that make it possible. It's time to look for a "Lighthouse."

Chapter 3

# Find Something to Listen To (Legally)

BEFORE YOU EVEN THINK about pressing that Push-To-Talk button to broadcast your voice to the world, we need to prove your radio can actually hear the world around it. Right now, you are holding a piece of plastic and metal that feels full of potential, but if it stays completely silent, that potential quickly turns into frustration. We need to find a 'known-good,' continuous signal to validate that your hardware is actually working. We are going hunting for a lighthouse.

**What you'll accomplish:**

- Find and tune into "known-good" signals that broadcast continuously.
- Confirm your receive path works in the real world (not just listening to raw static).
- Build massive confidence in using VFO mode to explore the airwaves without the pressure of transmitting.

**What you need:**

- Your Baofeng radio, powered on with the antenna attached.
- Your squelch set to Level 2 (from Chapter 2).
- A piece of paper and a pen.

## The Lighthouse Concept

When a sailor is navigating a dark, unfamiliar sea, they look for a lighthouse, a fixed, reliable, constant beam that proves their position.

When you are a beginner with a new radio, you need a lighthouse signal. If you just type in a random frequency and hear nothing, you have no idea if the frequency is dead, if you are out of range, or if your radio is broken. We need to find signals that we *know* are broadcasting right now.

## Target 1: The National Weather Service (NOAA)

For users in North America, the absolute best lighthouse is the NOAA Weather Radio system. These stations broadcast automated weather reports 24 hours a day, 7 days a week. It is a continuous, non-stop loop of spoken audio. If you can tune into this, you immediately prove your antenna, receiver, and speaker are perfectly functional.

NOAA broadcasts on seven specific frequencies. You will use VFO mode to check them one by one until you hear the robotic weather voice.

**Step 1:** Look at your screen. Make sure you are in **VFO Mode** (you should see just frequency numbers, not small channel numbers on the right). If you see channel numbers, press the orange **VFO/MR** button once.

**Step 2:** Use the blue **A/B button** to make sure the tiny black arrow is pointing to the top line.

**Step 3:** Use the keypad to type in the first frequency: 1 6 2 . 4 0 0.

FIG 3

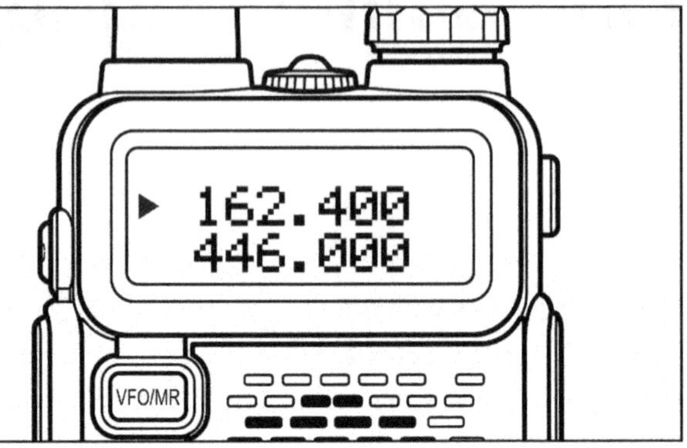


*Note: As you type the numbers, the radio will speak them out loud. When you hit the 6th number, it will load the frequency.*

Did the squelch open? Do you hear a voice reading a weather forecast?

- If **YES**, stop! You found your lighthouse. The voice will likely sound highly synthesized, repetitive, and robotic, this is completely normal.
- If **NO**, the station might be too far away. Type in the next frequency on the list:
  - 1 6 2 . 4 2 5
  - 1 6 2 . 4 5 0
  - 1 6 2 . 4 7 5
  - 1 6 2 . 5 0 0
  - 1 6 2 . 5 2 5
  - 1 6 2 . 5 5 0

*YOU'RE WINNING*

If you hear the weather broadcast, take a moment to celebrate. Your Baofeng is officially receiving real-world radio traffic! You can use the volume knob to dial in the perfect sound level.

## Target 2: Local Ham Radio Repeaters

If you aren't in North America, or if you live in a deep valley where weather radio doesn't reach, your next best lighthouse is a local amateur radio repeater.

Repeaters are large, powerful radio towers placed on tall buildings or mountains by local radio clubs. They listen for weak signals from handheld radios (like yours) and blast them back out at high power so everyone in the city can hear.

- **How to find them:** You will need to use your smartphone or computer. Go to a website like **RepeaterBook.com** or download their free mobile app. Search for your local city or county.

- **What to look for:** Look for repeaters operating in the "2 Meter" (144-148 MHz) or "70 Centimeter" (420-450 MHz) bands, as these are what your Baofeng can hear.
- **The Catch:** Unlike weather radio, repeaters are not continuous. You will only hear something if two people are actively having a conversation. You might have to type in the frequency (e.g., 1 4 6 . 5 2 0) and just let the radio sit on your desk for twenty or thirty minutes while you do other things. Don't stare at it. Just wait for the squelch to suddenly break and surprise you with a conversation.

## What Good Receive Sounds Like (vs. Bad Receive)

As you start exploring, you will hear a variety of different audio qualities. It takes a little practice to train your ear. Here is what to listen for:

- **Full Quieting (Excellent):** The person's voice comes through crystal clear, loud, and rich. There is almost zero background hiss. It sounds like they are standing in the room with you. This means you have a perfectly strong signal.
- **Scratchy/Picket-Fencing (Marginal):** You can hear the voice, but it is buried under a layer of static. If the person is driving a car while talking, the signal might rapidly chop in and out (this rhythmic chopping is called 'picket-fencing,' because it sounds exactly like listening to a car drive by while you stand behind a wooden picket fence).
- **Static Crash (Lost Signal):** You hear a sudden, loud burst of pure static, and then the squelch snaps the speaker shut again. The signal was strong enough to break the squelch gate for a split second, but too weak to hold it open.

*FIELD NOTE*

**Your body is part of the antenna system.** If you are listening to a scratchy, marginal signal, try standing up, walking near a window, or simply holding the radio higher in the air. UHF/VHF radio waves are strictly "line-of-sight." Even moving two feet to the left can completely clear up a bad signal.

## Exercise: The Receive Log

Grab your pen and paper. Today, we are just listening. Spend the next ten minutes typing in different frequencies from your local RepeaterBook search or the weather radio list.

Create a quick log with three columns:

1. **Frequency:** (e.g., 162.550)
2. **What I Heard:** (e.g., *Automated weather voice / Two older guys talking about antennas*)
3. **Signal Quality:** (e.g., *Crystal clear / Very scratchy*)

Writing this down does two things: it builds your muscle memory for typing on the VFO keypad, and it gives you a list of "known-good" frequencies that we are going to permanently save to your radio in Chapter 5.

**Quick Win Checklist**

- [ ] I have successfully switched my radio to VFO mode (manual typing).
- [ ] I understand why we hunt for a continuous "lighthouse" signal first.
- [ ] I have successfully tuned into a NOAA Weather Radio broadcast or a local repeater.
- [ ] I understand the difference between a "Full Quieting" signal and a scratchy signal.
- [ ] I have started a simple paper "Receive Log" and noted at least one working frequency.

## If you're stuck, go to...

- *Chapter 10: I Can't Hear Anything* (if you have tried all the weather frequencies and still hear only silence).

# Narrative Interlude: The Lighthouse Test

**CHARACTER:** NATE BROOKS (SOUTHEAST)

**Skills reinforced:**

- Known-good receive
- Confidence check
- Receive log
- "Listen-first" mindset

**Where to learn the steps:** Chapter 3

The humidity in the garage was a physical weight, the kind that signaled hurricane season was no longer a date on the calendar, but a looming reality. Nate Brooks stood at his workbench, surrounded by the orderly chaos of preparedness. He accounted for his mission-essential inventory: four-stage water filtration, primary and secondary illumination, and a trauma kit stocked for 72 hours of field care.

In the center of the bench sat the new Baofeng UV-5R.

Nate didn't have his ham license yet, that was a target for next month, but he was a man who believed in redundant systems. If the cell towers went down during a Category 3 surge, he needed a way to hear the outside world. He'd read the warnings about these radios being complicated, but he figured his years in the field and his disciplined approach to gear would see him through.

He clicked the radio on.

*"BEEP. FREQUENCY MODE."*

He kept the volume low. He didn't need his wife, Kara, coming out to the garage to hear the radio shouting at him. He'd promised her this wouldn't be another "project" that required a three-hour lecture to use. He needed it to be a tool: reliable, simple, and proven.

Nate began tapping the up-arrow key, scanning through the 144.000 MHz range. He heard a burst of static. Then a snippet of a voice, "...roger that, 73s..." then silence. He hopped to another frequency. More static. He tried scanning, but the radio seemed to skip over everything or stop on nothing but the rhythmic *chug-chug-chug* of local electrical interference.

Ten minutes in, Nate felt a flicker of uncharacteristic frustration.

"Is this it?" he muttered. He looked at the stock antenna. It was small, flexible, and felt... cheap. Maybe the reviews were right. Maybe he'd bought a thirty-dollar paperweight. If he couldn't hear a clear sentence from three miles away, how was he going to hear emergency coordinates when the sky turned purple and the power lines started snapping?

He almost reached for his phone to order a larger, "tactical" antenna, the kind that looked like it belonged on a tank. He had momentarily fallen into the gear-trap, assuming a hardware upgrade would compensate for a lack of verified procedure. He was searching for a single point of failure in the antenna when the real failure was in his diagnostic process. But then, he stopped. He looked at the "Ghostwriter Guide" pinned to the corkboard above his bench. He'd underlined a specific sentence in Chapter 3: *Before you can trust a tool, you need a repeatable test.*

He realized he was "frequency hopping" like a kid looking for a radio station on a road trip. He hadn't established a baseline. He hadn't found his lighthouse.

Nate sat down and pulled a fresh notebook toward him. At the top of the page, he wrote: **RECEIVE LOG.**

He checked the guide for his local "known-good" signals. The first was the National Weather Service. He knew the tower was located about twelve miles to the northwest. It was a massive, high-power broadcast. If the radio could hear anything, it would be that.

He typed the digits slowly: 1 6 2 4 0 0.

Silence.

His heart sank. "Nothing," he whispered.

Then he remembered the lesson from the previous chapter. He initiated a manual override of the squelch gate, holding down the **MONI** button to force the audio channel open and conduct a perimeter check of the signal threshold.

*SHHHHHHHHH,*

And underneath the hiss, a faint, rhythmic pulse.

Nate stood up and walked to the garage door. He hit the opener, and as the heavy metal door rumbled upward, the signal transformed. The static retreated, and a clear, robotic voice filled the garage.

*"...Highs in the lower nineties. Heat index values up to one hundred and five. A slight chance of thunderstorms in the evening..."*

"Target acquired," Nate said. He felt a surge of relief, but he didn't stop there. He went back to the bench and made his first entry in the log.

- **Freq:** 162.400 (NWS)
- **Location:** Garage (Door Open)
- **Quality:** 5/5 (Clear)
- **Note:** Signal drops inside the garage when the door is closed.

He moved to the next "lighthouse", the primary local ham repeater. He'd found the frequency on a local club's website: 146.760. He typed it in and waited.

For three minutes, there was nothing. Nate didn't touch the buttons. He didn't scan. He simply listened. He practiced the "listen-first" mindset. He knew the repeater was like a lighthouse; it didn't shine all the time, only when someone was using it.

Finally, the radio chirped.

*"K4XYZ monitoring,"* a voice said. It was crisp, loud, and sounded like the person was standing right in the driveway.

Nate checked the signal bars on the screen. Full strength. He didn't transmit. He couldn't, he wasn't licensed, and he respected the discipline of the airwaves. But he'd proven the most important part of the chain: **The Receive.**

He logged the second entry:

- **Freq:** 146.760 (Charleston Repeater)
- **Location:** Garage Workbench
- **Quality:** 5/5
- **Note:** High-traffic tower. Reliable for local info.

By the time Kara walked out to the garage twenty minutes later, Nate wasn't button-mashing or looking frustrated. He was calmly writing in his notebook.

"How's the new toy?" she asked, leaning against the doorframe.

"It's not a toy, Kara. It's a verified system," Nate said, turning the volume down. "I've established a receive baseline. I know exactly where to go for weather and exactly where to go for neighborhood status. I've documented the dead zones in the house."

Kara looked at the notebook, then at the radio. "You actually made it work? No lectures?"

"No lectures," Nate promised. "Just data. I know the radio works because I found the lighthouses."

Nate felt a new level of confidence as he closed the garage door. He didn't need a bigger antenna yet. He didn't need more gadgets. He had a receive chain he could trust because he'd bothered to test it before he needed it.

He looked at the radio sitting on the bench. It wasn't a "paperweight" anymore. It was a window into the local world.

"Now," Nate said to himself, looking at the next section of the guide. "If I can hear them, I need to understand why they might not be able to hear me. It's time to solve the offset puzzle."

## Takeaway

A radio is a tool of perception before it is a tool of transmission. Finding a "lighthouse signal" (like the NWS or a high-power local repeater) proves your hardware is functional and your settings are correct. Logging these successes creates a "map of certainty" you can rely on when a real crisis strikes.

**Bridge:** Nate has confirmed his radio is a capable listener. But a listener is only half of a conversation. To reach out and touch those distant towers, he has to master the math of the Repeater Puzzle.

Chapter 4

# The Repeater Puzzle Made Simple (Offset + Tone in Plain English)

You are tuning around in VFO mode, and suddenly, you hear a booming, crystal-clear voice. You've found a local repeater! Excited, you press the Push-To-Talk button to say hello, but the person keeps talking as if you don't even exist. You try again. Nothing. You pull away from the radio and wonder: *"I can hear the repeater... why can't it hear me?"*

**What you'll accomplish:**

- Understand "Offsets" with a simple, visual mental model (no physics degree required).
- Understand what a "Tone" is, what it isn't, and why it acts like a bouncer at a club.
- Use a dead-simple, 60-second decision chart to know exactly what to type in.

**What you need:**

- Your Baofeng radio.
- A listing for a local repeater (from an app like RepeaterBook or a local club website).

## The Repeater in One Diagram

To understand why the repeater can't hear you, you have to understand how repeaters actually work.

Imagine two people standing on opposite sides of a massive mountain. They have small handheld radios, but the radio waves smash directly into the rock. They cannot talk to each other.

Now, imagine we build a giant tower on the very top of the mountain. This tower has two things: a massive **microphone** pointing down at the city, and a massive **speaker** blasting audio out across the entire valley.

If the tower's microphone and speaker were on the *exact same frequency* at the *exact same time*, it would instantly create a deafening feedback loop, like holding a microphone too close to a concert speaker. To prevent this, the repeater uses **two different lanes on the highway.**

- It **listens** for your voice on one frequency (the uplink).
- It **transmits** your voice out to the city on a different frequency (the downlink).

FIG 4: THE REPEATER PUZZLE (HOW IT WORKS)

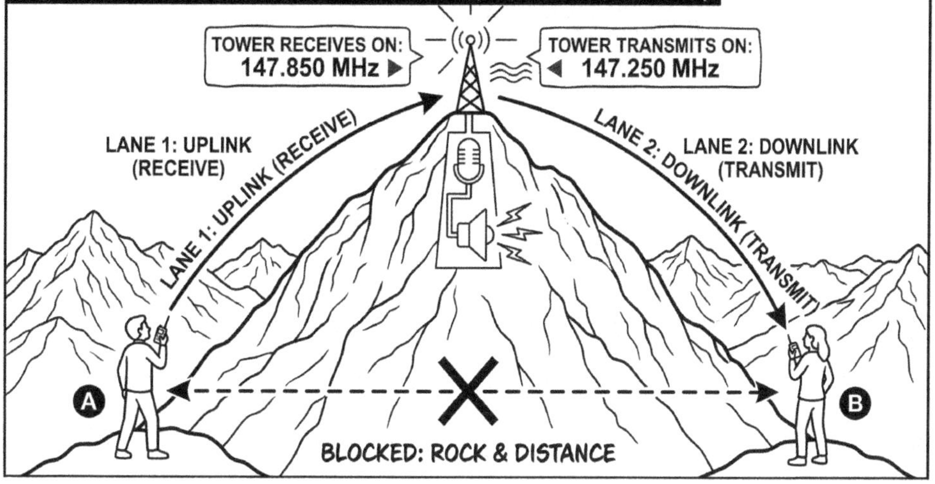


When you typed a frequency into your radio in Chapter 3, you typed in the repeater's *speaker* frequency. That's why you can hear it! But when you press the button to talk, your radio is transmitting on that same frequency. The repeater's microphone is tuned to a completely different channel, so it never hears you.

## Offset: The Math Between the Lanes

To fix this, we have to tell your Baofeng to automatically jump to the repeater's listening lane whenever you press the PTT button. This jump is called the **Offset**.

There are two parts to an Offset:

1. **The Direction (Duplex):** Does the radio need to jump *Up* (+) or *Down* (-) from the listening frequency?
2. **The Amount:** How far does it need to jump? (Usually 0.600 MHz for the 2-meter band, or 5.000 MHz for the 70-centimeter band).

*Example:* A repeater is listed at 146.700 with a Negative (-) offset of 0.600. When you listen, your radio stays on 146.700. When you press PTT, your radio does the math instantly and jumps down to 146.100 to talk to the tower. When you let go, it jumps right back to listen. *Don't panic, you do not have to do this math in your head. You just program the 0.600 once, and the radio handles the jumping automatically.*

## Tones: The VIP Password

Let's say you perfectly programmed the offset. You press the button, and the repeater *still* ignores you. What gives? You are missing the VIP password.

In busy cities, there is a lot of random static, interference, and overlapping radio traffic. Repeaters don't want to accidentally broadcast garbage to the whole city. So, they put an invisible "bouncer" at the door.

This bouncer demands a password. In the radio world, this password is a sub-audible tone (a low-pitched hum that your ear can't hear, but the computer can).

- These are officially called **CTCSS** (Continuous Tone-Coded Squelch System) or **DCS** (Digital Coded Squelch).
    - Many older hams just call them 'PL Tones' (which stands for Private Line, an old Motorola trademark that unfortunately makes beginners think it's a private, encrypted channel).

If your radio transmits your voice *without* adding this invisible tone to the background, the repeater's bouncer refuses to open the door, and your transmission hits a brick wall.

*FIELD NOTE*

**Tones are NOT encryption.** They do not hide your voice or make your transmission private. Anyone tuning to that frequency can still hear every word you say. They are purely a password to wake up the repeater tower. In the next chapter, we will show you exactly which keypad buttons to press to hand the bouncer your password.

**The 60-Second Decision Chart**

When you look at a repeater listing online, it can look like a spreadsheet of confusing numbers. Use this simple chart to strip away the noise and focus only on what matters:

**1. What is the Receive (RX) Frequency?**

- *Action:* Write this down. This is your baseline number.

**2. Does the listing show an Offset (+ or -) and an amount?**

- *Action:* Write down the direction (+ or -) and the number (like 0.600).

**3. Does the listing show an "Uplink Tone" (sometimes just called Tone or CTCSS)?**

- *If YES:* Write down the number (e.g., 100.0). You will need to program this into the T-CTCS menu.
- *If NO:* **Leave your tone off.**

*STOP*

**The Tone-Squelch Trap.** Do not touch the menu settings labeled R-CTCS (Receive Tone) or Tone Squelch. If you set a receive tone, you are putting a bouncer on *your own radio.* You will block out the repeater entirely and hear nothing but silence. You only ever want to set a **Transmit Tone** (T-CTCS) so you can open the repeater's door. Keep your own door wide open.

## Quick Win Checklist

- [ ] I understand that repeaters listen on one frequency and transmit on another to avoid feedback loops.
- [ ] I understand that the "Offset" is simply the math my radio does to jump between those two lanes.
- [ ] I know that a CTCSS/PL Tone is just an invisible password required to wake up the repeater tower.
- [ ] I know that I should only ever program a Transmit Tone, never a Receive Tone (unless explicitly needed).
- [ ] I have looked up one local repeater and written down its Receive Frequency, Offset, and Tone.

## If you're stuck, go to...

- *Chapter 9: I Can Hear Them, But They Can't Hear Me* (if you suspect you have the tone wrong).
- *Chapter 11: I Can Hit One Repeater but Not Another* (if you are confused by DCS vs CTCSS tones).

Chapter 5

# Program Your First Repeater (Keypad Method)

YOU UNDERSTAND HOW THE repeater works. You have the frequency, the offset, and the tone written down on a piece of paper in front of you. Now, you have to actually put them into the radio so they stay there.

This is the exact moment most people quit. The Baofeng menu system was designed by engineers, for engineers. It is incredibly unforgiving, and the specific sequence in which you press the buttons matters deeply. If you mess up the order, the radio won't give you an error message, it will silently pretend to save it, leaving you with a broken, frustrating channel.

Don't quit. We have the exact, step-by-step formula. Do not skip a single step.

**What you'll accomplish:**

- Enter the repeater receive frequency manually.
- Program the offset and tone correctly using the menu.
- Save the entire package to a permanent memory channel so it actually sticks.

**What you need:**

- Your Baofeng radio.
- Your piece of paper from Chapter 4 with the Repeater Frequency, Offset (+ or -), Offset Amount, and Transmit Tone.

## Step 1: Prepare the Canvas (VFO Mode)

You cannot program a channel while you are in Memory mode. You must be in the "manual typing" mode.

1. Press the orange **VFO/MR** button until your screen shows raw frequency numbers without any small channel numbers on the right edge.
2. Press the blue **A/B** button until the little black triangle arrow is pointing to the **TOP line** of the screen. (Always program on the top line to avoid glitching the radio).

## Step 2: Delete the Target Memory Channel

The Baofeng is stubborn. If you try to save a new frequency into Channel 001, but Channel 001 already has something in it from the factory, the radio will silently reject your save. You *must* sweep the floor before you build the house. Let's delete Channel 001 to make room.

1. Press **MENU**.
2. Type **2 8**. The screen will say DEL-CH (Delete Channel).
3. Press **MENU** to unlock it. The arrow drops down.
4. Use the Up/Down arrow keys (or type 0 0 1) to select Channel 001.
5. Press **MENU** to confirm the deletion. The voice will say "Confirm."
6. Press **EXIT**. The channel is now empty and ready for our data.

## Step 3: Type the Receive Frequency

On the keypad, type in the main listening frequency of your repeater. *(Example: 1 4 6 . 7 0 0)*

## Step 4: Set the Offset Direction

We need to tell the radio which way to jump when you press the PTT button.

1. Press **MENU**.
2. Type **2 5**. The screen will say SFT-D (Shift Direction).
3. Press **MENU** to unlock.
4. Use the arrow keys to select + (Plus), - (Minus), or **OFF** (based on your notes).
5. Press **MENU** to confirm.

## Step 5: Set the Offset Amount

Now we tell it *how far* to jump.

1. While still in the menu, press the **Up Arrow** once to go to Menu **2 6**. The screen will say OFFSET.
2. Press **MENU** to unlock.
3. Type the amount using exactly six digits. The Baofeng is not a smart device; if you just type '6 0 0', it will get confused. You must feed it all the leading zeros. *(Example: For a 0.600 MHz shift, type 0 0 0 6 0 0. For a 5.000 MHz shift, type 0 0 5 0 0 0).*
4. Press **MENU** to confirm.

## Step 6: Set the Transmit Tone (The Password)

If your repeater requires a tone to open the door, we set it here. If your repeater does *not* require a tone, you can skip this step entirely!

1. Type **1 3** while in the menu. The screen will say T-CTCS (Transmit CTCSS).
2. Press **MENU** to unlock.
3. Use the arrow keys to find your exact tone number *(Example: 1 0 0 . 0).*
4. Press **MENU** to confirm.
5. Press **EXIT** to leave the menu system.

## Step 7: The Final Save (Crucial Sequence)

You have successfully built the profile in VFO mode. Now, we have to lock it into the permanent memory slot (Channel 001) that we cleared earlier.

1. Press **MENU**.
2. Type **2** 7. The screen will say MEM-CH (Memory Channel).
3. Press **MENU** to unlock.
4. Use the arrow keys (or type 0 0 1) to select Channel 001.
5. Press **MENU**. The robotic voice will say **"Receiving Memory."**

FIG 5

*STOP*

**Do not press Exit yet!** The Baofeng just saved the *listening* frequency, but it hasn't locked in the *talking* frequency and offset math yet. We have to save it a second time.

1. Press **MENU** again. The arrow drops down.
2. Make sure Channel 001 is still selected.
3. Press **MENU** one more time. The voice will say **"Transmitting Memory."**
4. Now, press **EXIT**.

*YOU'RE WINNING*

You did it! Press the orange **VFO/MR** button to switch over to Memory Mode. Navigate to Channel 001. You should see your repeater frequency. When you press the PTT button, you should see the frequency on the screen physically jump to the new talking lane, and jump back when you release it. You have successfully programmed a Baofeng manually!

## Quick Win Checklist

- [ ] I always clear/delete a memory channel before trying to save new data into it.
- [ ] I successfully typed my repeater frequency into VFO mode.
- [ ] I set my Offset Direction (Menu 25) and Offset Amount (Menu 26).
- [ ] I set my Transmit Tone (Menu 13) if the repeater requires one.
- [ ] I followed the double-save sequence in Menu 27, hearing "Receiving Memory" and then "Transmitting Memory."
- [ ] I have switched back to Memory Mode and tested the channel recall.

## If you're stuck, go to...

- *Chapter 12: My Channels Won't Save* (if you switch to Memory Mode and the channel is blank or behaves strangely).

# Narrative Interlude: Why the Repeater Hears You, Or Doesn't

### Offset for Humans

**Character:** Mark Jensen (Midwest Suburbs)

**Skills reinforced:**

- Offset / Duplex understanding
- Transmit Tone vs. Tone-Squelch
- Keypad Programming
- The "Double Save" Order

**Where to learn the steps:** Chapters 4 & 5

Mark Jensen stood at the end of his driveway, bathing in the crisp Minnesota evening air, standing as tall as a suburban sentinel. He was convinced he was exactly one minute away from greatness.

In his right hand, he held the Baofeng UV-5R. He was officially a licensed ham. He had conquered the squelch wall in his kitchen two days ago, and now, he was ready to graduate from merely eavesdropping on the automated weather robot to talking to actual human beings on the primary Twin Cities repeater.

He had his cheat sheet in his left hand. *Receive Frequency: 146.700. Minus offset. Tone 114.8.*

"Okay. This is... a lot of numbers," Mark muttered to himself. "But I am a man of science and precision."

He punched 1 4 6 7 0 0 into the keypad. Instantly, the radio crackled to life. Two guys were discussing the best type of coax cable to run through a dry-walled garage ceiling. Mark grinned. He was at the door of the club. Now he just needed to walk in.

He pressed the Push-To-Talk button. He took a breath, ready to announce his callsign to the world.

Nothing happened. The two guys just kept talking right over him.

Mark frowned. He tried again, holding the radio closer to his face like a stubborn slice of pizza. Still nothing. The radio didn't even make that polite, secondary *beep* the book said repeaters make when they hear you.

"Right, the password," Mark said out loud. He dove into the menus. He remembered seeing something about 'Tones'. He aggressively pressed the menu button, scrolled until he found R-CTCS, and proudly typed in 114.8. *Receive Tone. Obviously,*

Mark thought. *I want to receive them, don't I? Perfect.* Then he remembered the offset thing. He went to OFFSET and typed 000600.

He hit Menu 27 to save it to Channel 001, heard the robotic voice say *"Receiving Memory,"* and excitedly pressed Exit.

He switched to Memory Mode, pulled the radio to his mouth, and pressed PTT.

Silence. Not only was the repeater ignoring him, but the two guys talking about coax cable had completely vanished. The radio was as dead as a brick.

"I touched one button and now I'm in another universe," Mark groaned, letting his arms flop to his sides. "I've broken it. I've been banned from the airwaves before I even said hello."

"Having a fight with the tactical walkie-talkie?"

Mark jumped, feeling like a teenager caught playing with a toy lightsaber. His neighbor, Ron, was leaning over the property line fence, holding a pair of garden clippers. Ron was in his late sixties, a retired electrical engineer, and possessed an aura of relentless, practical calm. More importantly, Ron was an Extra Class ham radio operator. He had an antenna on his roof that looked like a metallic spider web.

"Ron. Hey," Mark said, trying to casually hide the Baofeng behind his leg. "Just... testing propagation. Atmospheric stuff."

"Right," Ron chuckled, walking over to the edge of Mark's driveway. "Let me guess. You can hear the repeater, but it acts like you don't exist. So you went into the menus, changed a bunch of things, and now you can't hear anything at all."

Mark blinked. "Are you psychic?"

"I'm experienced," Ron smiled. "I heard the Baofeng error beeps from my tomato garden. Hand it here."

Mark handed over the radio. Ron adjusted his glasses, glanced at the screen for half a second, and immediately shook his head. "Well, first off, you put a bouncer on your own door, Mark."

"A what?"

"Menu 11. R-CTCS. You set a receive tone. That tells your radio to ignore everything in the world unless the incoming signal has a secret password. The repeater doesn't send passwords back to you, so your radio just locked itself inside a soundproof vault." Ron pressed a few buttons with blazing speed. "Turned that off. Keep your ears open. Now, let's look at your transmit side."

"I did the offset," Mark protested defensively. "Six hundred. And I saved it."

"Let's gather the four numbers first," Ron said patiently, holding up four fingers. "No guessing. What's the receive frequency, the offset direction, the offset amount, and the transmit tone?"

Mark read them off his wrinkled cheat sheet.

"Good. Now, here is the secret of the repeater," Ron explained, gesturing with his garden clippers. "It's a two-lane highway. You listen on the southbound lane, but you have to talk on the northbound lane. If you just shout into the southbound lane, you're going the wrong way, and the tower's microphone never hears you."

Mark nodded slowly. "So the offset is just... the radio changing lanes?"

"Exactly," Ron said. "But the Baofeng is a very stubborn student. You have to teach it exactly how to change lanes, and you have to save it in a very specific order."

Ron handed the radio back to Mark. "Do it with me. Let's sweep the floor first. Menu 28. Delete Channel 001. Get rid of the corrupted file you just made."

Mark followed the steps. *Delete Channel. Confirm.*

"Now," Ron instructed. "Top line. VFO mode. Type the receive frequency."

Mark typed 1 4 6 7 0 0.

"Menu 25. Shift Direction. Make it a minus." Mark did so. "Menu 26. Offset. Type 000600." Mark confirmed it. "Menu 13. Transmit Tone, or T-CTCS. This is your VIP password for the tower. Set it to 114.8."

"Done," Mark said, his thumbs starting to sweat.

"Now, the trapdoor," Ron said, stepping back and folding his arms. "Menu 27. You have to do the double-save. If you just save it once and exit, the radio only remembers the listening lane. You have to save the talking lane, too."

Mark went to Menu 27. He selected Channel 001. He pressed Menu. *"Receiving Memory,"* the robot voice chimed. "Don't touch Exit," Ron warned. "Press Menu again." Mark pressed it. The arrow dropped. He pressed it a third time. *"Transmitting Memory."*

"Boom," Ron said. "You just saved both lanes. Exit out. Switch to Memory mode. Try it."

Mark pressed the orange VFO/MR button. The tiny 001 appeared on the right. He pulled the radio up to his mouth. He pressed the PTT button, but before he could speak, he looked at the screen. The numbers had instantly jumped from 146.700 down to 146.100.

He let go of the button. The numbers jumped right back to 146.700. Then, the radio let out a crisp, polite *beep* from the speaker, followed by a tiny burst of static.

"The repeater tail," Ron grinned. "The tower heard you, opened the door, and closed it again."

"Look at that," Mark breathed, staring at the plastic device with newfound reverence. "It actually worked. The math actually worked."

"Repeaters aren't hard, Mark," Ron said, tapping the Baofeng's antenna lightly with his clippers before turning back toward his yard. "Guessing is hard. Stop guessing. Get your four numbers, check your bouncers, and follow the exact saving order. You'll be fine."

Mark stood in the driveway alone again. The sun was setting, casting a golden hour glow over the suburbs. He had done it. He had a fully functioning, programmed repeater saved to Channel 1. He wasn't locked out anymore.

He looked down at the PTT button. The technical hurdle was gone. The only thing standing between him and the rest of the amateur radio world was his own voice.

"Okay," Mark said aloud to the empty driveway. "This is the part where I pretend I'm not terrified of pressing the button."

## Takeaway

Repeaters aren't hard, guessing is hard. Never set a Receive Tone (R-CTCS) unless you absolutely have to, or you will lock yourself out of hearing traffic. Gather your four pieces of data (RX Freq, Offset Direction, Offset Amount, TX Tone) and always finish with the "Double-Save" to ensure both the listening and talking lanes are locked into memory.

**Bridge:** Mark's radio is programmed perfectly. Now comes the scary part: pressing PTT and speaking to the world for the first time.

Chapter 6

# Your First On-Air Moment (What to Say + What NOT to Do)

You've programmed the repeater perfectly. You switch over to Memory Mode, select Channel 1, and the screen looks exactly right. Now, you're staring at the Push-To-Talk (PTT) button like it's the launch button for a nuclear missile.

"Mike fright" is incredibly common. The thought of broadcasting your voice to a whole city can cause your brain to freeze. You worry about sounding foolish, breaking a rule, or using the wrong jargon. Take a deep breath. Every single expert on the radio today was terrified the first time they pressed that button. Think of this like using an intercom at a hardware store to ask a quick question, not giving a presidential address. Nobody expects you to sound like a professional broadcaster. We are going to keep it short, polite, and completely painless.

**What you'll accomplish:**

- Learn the absolute simplest, stress-free script for your first legal transmission.
- Understand the 5 golden rules of radio etiquette that prevent awkwardness.
- Understand basic identification rules so you stay on the right side of the law.

**What you need:**

- Your Baofeng radio, tuned to your newly saved repeater channel.
- An Amateur Radio License (Required before you press the PTT button!).
- Your official Call Sign.

## The 5 Golden Rules of Etiquette

Before we talk, we need to know how to behave. Ham radio is a community, and like any community, there are unwritten rules of politeness.

1. **Listen First:** Never turn on a radio and immediately start talking. The channel might be in the middle of a serious conversation, an emergency, or a weekly club meeting. Wait at least 30 to 60 seconds to ensure the airwaves are clear.
2. **Press, Pause, Speak:** When you press the PTT button, do not start speaking immediately. It takes about a half-second for the repeater tower's computer to hear your invisible tone, verify your password, and physically open its gates. Think of it like walking up to the automatic doors at a grocery store, you have to pause for a beat to let them open before you walk through. If you speak instantly, your first word will bounce off the closed door and be cut off. Press the button, take a quick breath, and *then* speak.
3. **No "CQ" on Repeaters:** You might have seen movies where radio operators yell "CQ, CQ, CQ" (calling any station). This is for long-distance, static-filled global radio. On a crisp, local city repeater, yelling CQ sounds bizarre. We just use plain English.
4. **Don't Overstay Your Welcome:** Repeaters have automatic timers built-in. If you hold the button down and talk non-stop for 3 minutes, the repeater will automatically shut off (time-out) to prevent overheating. Keep your transmissions to 30 seconds or less. Talk, then listen.
5. **No Profanity:** The airwaves are public, monitored by the FCC (or local equivalents), and families are often listening. Keep it PG-rated.

## The Simplest Script: "Monitoring"

You do not need a reason to talk on a repeater. You don't need a specific topic. The most common transmission in ham radio is simply letting people know you are alive, your radio works, and you are listening if anyone wants to chat.

Here is the exact script. Just replace the bracketed text with your official call sign (e.g., K-1-A-B-C).

*(Press PTT... wait half a second...)* "This is *Your Call Sign*, monitoring." *(Release PTT)*

That's it. Two words. You have officially made your first transmission. If the repeater heard you, you will usually hear a short burst of static, or a polite "beep" from the tower acknowledging you, when you release the button.

If someone is out there driving in their car, they might reply: "Hey *Your Call Sign*, this is *Their Call Sign*. Good morning, how are you doing today?" If they do, just reply normally! "Doing great, just testing out a new Baofeng radio. Thanks for coming back to me."

**Alternative Script: The Radio Check**

If you specifically want to know how your audio sounds, you can ask for a "signal report." Hams love giving signal reports; it makes them feel helpful.

*(Press PTT... wait half a second...)* "This is *Your Call Sign*, testing a new radio. Can anyone give me a quick signal report?" *(Release PTT)*

Someone will likely come back and say something like, *"You are loud and clear, full quieting,"* or *"You are a little scratchy, maybe step outside."* Thank them, give your call sign one more time to close out the conversation, and you are done.

## Practice: The Mirror Drill

Before you press the button for real, practice out loud in an empty room.

*FIELD NOTE*

**Phonetic Alphabet:** Say your call sign phonetically if you can (e.g., Kilo-One-Alpha-Bravo-Charlie). Phonetics are used because letters like 'B', 'C', 'D', and 'E' all sound identical over static. However, plain letters are perfectly fine for a beginner.

Hold the radio about two inches from your mouth. Speak *across* the face plate of the microphone rather than pointing it directly into your lips like a rockstar, this prevents your breath from 'popping' loudly on the audio. Speak in your normal, calm, indoor voice. Do not yell. Say the script out loud three times. Muscle memory defeats anxiety every single time.

## Quick Win Checklist

- [ ] I understand that I must possess a valid amateur radio license to legally perform this step.
- [ ] I have listened to my chosen channel for at least 60 seconds to ensure it is clear.
- [ ] I understand the "Press, Pause, Speak" rule to prevent my first words from being cut off.
- [ ] I have practiced my short script ("This is *Call Sign*, monitoring") out loud in an empty room.
- [ ] I know not to yell "CQ" on a local FM repeater.
- [ ] I have successfully pressed the PTT button and made my first on-air transmission!

## If you're stuck, go to...

- *Chapter 9: I Can Hear Them, But They Can't Hear Me* (if you made your call, but nobody replied, and you didn't hear the polite "beep" from the repeater).

Chapter 7

# Save a Clean “Home Channel Plan” (5 Channels You’ll Actually Use)

You’ve successfully programmed one repeater and made your very first transmission. Take a moment to appreciate that. You are lightyears ahead of most people who buy this radio, turn it on, hear static, and throw it into a closet in frustration.

But having just one working channel isn't quite enough for real preparedness, and falling into the trap of having 128 random, unnamed frequencies you downloaded from the internet is overwhelming. A radio full of mystery channels isn't a communication tool; it's just a noisemaker.

If an actual emergency happens, a severe winter storm knocks out regional power, or cellular networks go down due to overload, you do not want to be fumbling with VFO mode, offset math, and confusing menu settings in the dark. You need a small, highly organized, hyper-reliable set of channels you trust with your life. You need to build a "Home Profile."

**What you’ll accomplish:**

- Understand the psychology behind the 5-channel "Home Profile" and why less is actually more.
- Organize your memory channels logically so you can find exactly what you need in the dark, under stress.
- Create a physical, weatherproof paper backup so your family can use the radio even if you aren't around to help them.

**What you need:**

- Your Baofeng radio.
- A small index card or heavy piece of paper.
- A thick permanent marker.
- Clear packing tape (to laminate and waterproof your card).

## The "Top 5" Philosophy: Defeating Cognitive Overload

Your Baofeng holds up to 128 distinct channels in its memory bank. When new owners discover this, the immediate temptation is to plug the radio into a computer and fill every single slot with police scanners, fire dispatch, ambulance channels, and distant repeaters from three counties away.

Don't do it. Treat your radio like a neatly organized toolbox, not a junk drawer.

When human beings are under severe stress, like during a natural disaster or a medical emergency, our brains drop complex tasks. Fine motor skills degrade, your heart rate spikes, and tunnel vision sets in. If you have to desperately scroll through 90 channels of chaotic static and irrelevant police dispatch to find the *one* frequency your family uses to talk, the radio has become a liability, not an asset.

We highly recommend saving your 5 most important, mission-critical frequencies into Memory Channels 001 through 005. Think of these exactly like the emergency speed-dial favorites on your smartphone. You want to be able to turn the radio on and click up exactly three times, purely by muscle memory, knowing exactly who will be on the other end.

## The 5 Channels You Actually Need

Here is the perfect, stress-proof template for a beginner's Home Profile:

**Channel 001: The Primary Local Repeater** This is the reliable "Lighthouse" we found and programmed in Chapters 3 and 5. This is your daily driver. It is your most powerful tool for reaching out across the city to get critical information, ask for assistance, or make contact with local emergency nets.

**Channel 002: The Backup Repeater** Radio towers lose power, too. Backup generators run out of diesel, and heavy winds can snap antennas. You need redundancy. Jump back on RepeaterBook and find a second repeater, preferably one located on a completely different mountain or the opposite side of town, and program it exactly like you did the first one. If Channel 1 goes dead, you immediately click over to Channel 2.

**Channel 003: Simplex 1 (Primary Family Channel)** "Simplex" means pure, radio-to-radio direct communication without using a repeater tower as a middleman. It is strictly short-range (usually 1 to 3 miles in a suburban neighborhood), but it is the ultimate backup because it relies on zero external infrastructure. It works even if the entire city grid collapses. Pick a standard 2-meter simplex frequency (like 146.520, the national calling frequency, or 146.415) and save it. No offsets, no tones. Just your radio talking directly to their radio.

**Channel 004: Simplex 2 (Backup Family Channel)** If Channel 3 is suddenly busy with other people talking, you need a pre-planned alternate lane to switch your family over to. Pick a frequency in the 70-centimeter band (like 446.000) and save it. Having one 2-meter backup and one 70-centimeter backup gives you options if one frequency band is acting up due to local interference.

**Channel 005: Local Weather Radio (Receive Only)** The automated NOAA weather broadcast you found in Chapter 3. Save this to Channel 5. *Important note: The Baofeng physically cannot transmit on weather frequencies, but having it in your top 5 quick-dial means you can instantly get emergency weather updates with a single click, without having to type numbers in VFO mode.*

## Memory Layout Strategy (The A/B Defensive Setup)

Because the Baofeng has a top line (A) and a bottom line (B) on the display screen, you have the unique ability to actively monitor two of these vital channels at the exact same time.

A fantastic defensive posture for a storm or regional power outage is to set the Top Line (A) to **Channel 001** (your primary city repeater to listen for wide-area news and emergency nets) and set the Bottom Line (B) to **Channel 003** (your family's short-range simplex channel).

The radio will rapidly scan both simultaneously. If the city tower speaks, you hear it. But if your spouse calls you from down the street on the simplex channel to say they are walking home safely, the radio will instantly switch down to the bottom line to let you answer them.

## The Physical Backup (The "Hit By A Truck" Rule)

Electronics fail. Batteries die in the cold. Memory banks occasionally get accidentally wiped by a clumsy thumb pressing the wrong menu button in a dark tent. If your radio resets to the factory defaults during a crisis, all your hard work is instantly gone. Furthermore, if you are injured or away from the house, your family members need to know how to use the radio without reading this entire book.

Take a small, heavy index card. Write your 5-Channel Home Profile on it in dark, bold permanent marker so it is highly visible.

- **CH 1:** Primary RPT (146.700 / - / Tone 100.0)
- **CH 2:** Backup RPT (444.100 / + / Tone 123.0)
- **CH 3:** Family Direct 1 (146.415 / No Tone)
- **CH 4:** Family Direct 2 (446.000 / No Tone)
- **CH 5:** Weather Alert (162.550)

Cover both sides of this index card entirely in clear packing tape to make it waterproof and tear-resistant. Tape this card directly to the back of the radio (right over the battery block), or fold it up and place it prominently inside the padded case you keep the radio in. If someone else needs to pick up your radio, they can look at this card, look at the screen, and instantly understand exactly how to reach you.

## Quick Win Checklist

- [ ] I understand why a small, trusted "Home Profile" is vastly superior to a cluttered 128-channel list during a crisis.
- [ ] I have successfully programmed my Primary and Backup repeaters into Channels 1 and 2.
- [ ] I have successfully programmed two direct "Simplex" channels into Channels 3 and 4 for pure, short-range family use.
- [ ] I have successfully programmed a receive-only Weather frequency into Channel 5.
- [ ] I understand how to use the A/B dual-display to monitor both the city repeater and my family channel at the same time.
- [ ] I have written my 5-channel plan down on a physical index card, waterproofed it, and attached it to my radio kit.
- [ ] I have completed Part 1! I can officially operate a Baofeng radio with confidence!

## If you're stuck, go to...

- *Chapter 15: Reset and Recover* (if you accidentally filled your memory channels with junk, got overwhelmed, and want to completely wipe the radio clean to start fresh).

# Narrative Interlude: Thumb Off the Trigger

**Character:** Nate Brooks (Southeast)

**Skills reinforced:**

- First-call script (practice)
- Etiquette (listen-first mindset)
- The 5-channel Home Profile
- Creating a paper backup

**Where to learn the steps:** Chapters 6 & 7

The thick South Carolina heat had finally broken, leaving behind a muggy, cricket-filled evening that smelled faintly of distant thunderstorms and wet asphalt. Nate Brooks sat at his garage workbench, the overhead LED lights casting a stark, surgical glow over his neatly arranged gear. Everything had its place: the water filtration kits on the bottom shelf, the medical go-bags on the middle, and his communications array taking center stage. In front of him, the Baofeng UV-5R stood upright in its charging cradle base, its green screen illuminated as it quietly scanned the local frequencies.

He had spent the last hour meticulously following the programming guide. He had successfully locked the primary Charleston repeater into Channel 1, complete with the correct minus offset and the exact DCS tone required to digitally open the tower's squelch gate. He had done the math. He had swept the memory floor clean of junk frequencies and executed the double-save precisely as instructed. The infrastructure was built.

Suddenly, the radio broke the silence with a crisp burst of static, followed by a calm, authoritative voice that filled the garage.

*"This is W4ABC, Net Control for the Tri-County Skywarn training net. We are now taking check-ins from mobile and portable stations. Please call now."*

Nate leaned forward, resting his forearms on the workbench. He could hear a slight pause, then another voice chimed in, clear and professional.

*"W4ABC, this is Kilo-Four-Echo-Mike. Portable station, Mount Pleasant. Sky is overcast, no precipitation. Over."*

*"Acknowledged, K4EM. Thank you for the check-in,"* the Net Control operator replied warmly.

It was an orderly, disciplined roll call of people proving their equipment worked. There was no talking over one another, no CB-radio slang, and no panic. Just a methodical network of citizens creating a web of situational awareness across the county. It was exactly the kind of network Nate wanted to be a part of.

Nate picked up the Baofeng. The plastic casing felt solid and utilitarian in his palm. His thumb naturally found its way to the large, textured rubber Push-To-Talk (PTT) button on the left side of the chassis.

*Just one click,* a voice in his head rationalized. *Just give a fake name and ask for a quick radio check. How else are you going to know if the offset math actually worked? You need to verify the transmit chain before hurricane season officially starts. It'll take five seconds.*

He applied a fraction of an ounce of pressure to the PTT button. He could feel the mechanical switch beneath the rubber beginning to yield, right on the edge of breaking the squelch gate and broadcasting his voice to three counties.

Then, he stopped. He looked at the radio, then at the storm-prep bins stacked neatly against the garage wall. Each bin represented a commitment to doing things the right way, no shortcuts, no half-measures. He pulled his thumb completely off the trigger and set the radio back down on the workbench with a soft clack.

"Discipline," Nate muttered quietly to himself, staring at the green screen.

He wasn't licensed yet. He didn't have a call sign. Pressing that button right now wouldn't be a valid test of his gear; it would be a failure of his own operational discipline. The radio was just a machine, it would blindly do whatever he commanded. The real vulnerability in his preparedness plan wasn't the antenna, the battery, or the microchips. It was the operator. It was his own impatience. If he couldn't follow the basic, established legal rules of the airwaves on a perfectly calm Tuesday night, how could he trust himself to stay grounded and methodical when a Category 4 hurricane was churning off the coast and the cell networks were dead?

Nate took a deep breath, letting the tension leave his shoulders. The radio worked perfectly. He had proven the receive side beyond a shadow of a doubt. The transmit side would simply have to wait until he had earned the legal right to use it. But that didn't mean he couldn't finalize the architecture of his system tonight.

He opened his logbook notebook to a blank page. If an emergency happened while he was at work, he wouldn't be the only one needing to operate this radio. His wife, Kara, had made her boundaries incredibly clear when he first brought the box home: *One page. No menus. No lectures.* She was pragmatic. She wasn't going to memorize frequency offsets, tone squelch codes, or standard operating procedures. She needed a tool, not a new hobby.

Nate grabbed a bright yellow index card, chosen specifically for high visibility by flashlight, and a thick black permanent marker. He began to draft his "Home Profile." He needed a small, locked-down ecosystem of channels that required zero guesswork for anyone who picked it up.

In crisp, block letters, he wrote:

**CH 1: PRIMARY (Charleston City Repeater - City Info) CH 2: SECONDARY (North County Backup Repeater) CH 3: DIRECT 1 (Neighborhood Simplex - 1 Mile) CH 4: DIRECT 2 (Backup Simplex) CH 5: WEATHER (Receive Only - NWS)**

He reviewed the list, nodding in satisfaction. It was clean. It was redundant. It provided two avenues for long-range city intelligence, two avenues for short-range neighborhood security and family coordination, and an automated weather beacon. Five channels were all they realistically needed. Anything more would just become mental clutter during a crisis.

Nate pulled a roll of clear, heavy-duty packing tape from his tool drawer. He carefully laid strips of tape over the front and back of the yellow index card, pressing out the air bubbles with his thumb until it was completely sealed, waterproof, and tear-resistant. He punched a small hole in the top corner and used a short loop of high-visibility orange paracord to tie the card directly to the zipper of the padded radio pouch.

"Okay," Nate said, looking at the finished product. "The system is built and documented. Now we drill."

He picked up the radio again. He deliberately kept his thumb away from the PTT button this time. Instead, he looked at the blank garage wall, visualizing a real grid-down scenario. He practiced the etiquette of the "listen-first" mindset, counting a full *one-Mississippi, two-Mississippi* in his head to wait for the imaginary repeater tail to drop. Muscle memory, he knew, wasn't just for your hands; it was for your vocal cords, too.

He held the radio two inches from his mouth, angled slightly past his cheek to avoid popping his P's into the microphone, and spoke aloud in the empty garage.

"This is... well, pending call sign. Monitoring." He paused, feeling slightly foolish talking to a wall, but knowing the absolute necessity of vocal repetition under zero stress. "Let's try the check-in format. Who, Where, What, Next."

He cleared his throat and tried again, speaking with a steady, flat cadence designed to cut through static. "Kara, this is Nate. I am secure at the neighborhood entrance. Roads are clear of debris. I will check back on this channel at the top of the hour. Over."

"Loud and clear," a voice said from the doorway.

Nate turned. Kara was standing there, the condensation dripping down the sides of two glasses of iced tea she was holding. She walked over, set one down on the workbench, and let her eyes fall on the bright yellow, laminated index card tied to the radio pouch.

She picked it up, her eyes scanning the five channels.

"Channel Three to talk to you directly, Channel One to hear what the city is doing," she read aloud, tracing the heavy marker lines with her finger. She looked at him, raising an eyebrow in genuine surprise. "You actually distilled it down to five lines? No digging into menus? No three-hour lecture on radio wave physics?"

"No menus," Nate confirmed, taking a sip of the iced tea. "You turn the top knob on. You use the arrows to go to Channel 3. You press the side button and talk like a normal person. That's the entire system."

Kara nodded slowly, clearly approving of the brutal, elegant simplicity. "I can do that. It makes sense. Does it actually reach the city tower from here?"

"The receive does, I've confirmed that," Nate said. "I can't verify the transmit yet. Not legally, anyway."

He reached past the radio, opened his laptop on the corner of the bench, and quickly navigated to the local amateur radio club's website. He clicked on the testing schedule, found the session listed for that upcoming Saturday morning at the local community center, and typed in his credit card information to pay the fifteen-dollar exam fee.

"But I will be able to verify it by Saturday afternoon," Nate added, closing the laptop with a definitive snap.

He looked back at the radio. It was no longer a mystery box of confusing buttons or a temptation to break the rules. It was a fully prepped, documented, and organized communications node. It was ready for the exact moment he earned his call sign. There was no panic, and there was no guessing. He had built the plan, established the discipline, and locked in the procedure before he ever needed to press the button.

## Takeaway

Discipline beats impulse. You can build the system now, even before you transmit. By establishing a clean, 5-channel Home Profile and creating a laminated, physical backup card, you ensure your family can use the radio safely without needing a technical lecture. Practice your scripts out loud so muscle memory takes over when it counts.

**Bridge:** But if something breaks, I need a triage plan, not a spiral.

# Part 2:

Fix It Fast (Troubleshooting Tree)

Chapter 8

# Troubleshooting: How to Diagnose in 60 Seconds

You press the Push-To-Talk button, expect to hear a friendly reply, and instead... nothing happens. Or worse, you turn the radio on and it immediately screams a wall of deafening static at you. When your communication equipment fails, especially if you are outdoors, in the dark, or in the middle of a stressful situation, your heart rate naturally goes up.

Your first instinct is to assume you bought a defective piece of junk or that you broke it beyond repair. You don't want to read a twenty-page chapter on radio wave propagation theory or dig through a badly translated factory manual right now. You just want to know exactly which button to press to make the machine work again.

Welcome to the triage tent. Take a deep breath. In this section, we abandon deep theory entirely and focus strictly on rapid, symptom-based problem solving to get you back on the air in 60 seconds or less.

**What you'll accomplish:**

- Shift your mindset from "my radio is permanently broken" to "my radio just has a setting out of place."
- Master the "Big Three" checkpoints that solve 95% of all Baofeng problems in under a minute.
- Build massive confidence in your ability to fast-debug your own gear in the field without relying on the internet.

**What you need:**

- Your misbehaving Baofeng radio.
- A calm, methodical mindset.

## The Triage Mindset

When a car won't start, a good mechanic doesn't immediately assume the engine has exploded and the car is totaled. They check to see if there is gas in the tank, they check the battery connections, and they verify the key is actually turned.

Your Baofeng is a tough little industrial computer. It can survive being dropped in the dirt, left in a cold truck, or bumped around in a hiking backpack. It is highly unlikely that the internal circuit board is physically broken. Almost every single "failure" a beginner faces is simply a software setting that got bumped by a stray palm, changed by accident, or saved incorrectly in the confusing menu system.

If you can shift your mindset from "this radio is broken" to "a setting is out of place," you will save yourself hours of frustration. If we systematically check the core settings, we will find the cure every single time.

## The "Big Three" Checkpoints

Before you dive into the specific symptom chapters ahead, run your radio through this quick 60-second diagnostic flow. These three checkpoints solve the vast majority of user errors.

**Checkpoint 1: Are you in the right mode?** The Baofeng's split personality (VFO vs. Memory) is the number one cause of user lockouts. If you are trying to type a raw frequency manually using the keypad, but you see tiny channel numbers (like 001 or 015) on the far right edge of the screen, the radio is going to actively fight you. It will beep angrily and ignore your typing.

- *The Fix:* Press the orange **VFO/MR** button until the small channel numbers on the right side of the screen completely disappear. You are now safely in VFO mode. Remember: Memory mode is for securely listening to what you've already built; VFO mode is your blank canvas for making changes.

**Checkpoint 2: Is the black arrow pointing at the right line?** The Baofeng has a dual-watch display featuring a top line (A) and a bottom line (B). It is incredibly common to meticulously program a perfect frequency on the top line, setting your offsets and tones exactly right, but accidentally leave the active control arrow pointing to the bottom line. When you press Push-To-Talk, you are broadcasting on completely the wrong frequency. You feel like the radio is ignoring you, but it's actually doing exactly what the arrow told it to do.

- *The Fix:* Always look for the tiny black triangle on the left side of the screen. Before you press the PTT button, press the blue **A/B** button to physically move that little black arrow to the line you actually intend to use. Making this a habit prevents 50% of all transmission errors.

**Checkpoint 3: Is the battery actually seated?** If the radio is acting truly possessed, turning off randomly in your hand, the screen flickering or dimming when you press the PTT button to talk, or the radio completely restarting itself in the middle of a sentence, you do not have a software problem. You have a physical power delivery problem. Brand new Baofengs have famously stiff plastic battery rails, and it's easy to think the battery is locked when it's actually loose. A microscopic disconnect of the metal contacts will cause the radio's computer to instantly reboot and reset your active screen.

- *The Fix:* Remove the battery entirely by pressing the release latch. Wipe the metal contacts on both the battery and the back of the radio with a clean shirt to remove any factory grease or dust. Slide the battery back onto the rails, and press upward firmly with your thumbs until you hear a loud, undeniable *click*.

*YOU'RE WINNING*

If you can memorize these three checkpoints, you are officially the designated "radio tech" for your family or friend group. You will look like an absolute wizard to other beginners just by pressing the VFO/MR or A/B button to magically "fix" their broken radios.

## Quick Win Checklist

- [ ] I understand that 95% of radio problems are just bumped settings, not broken hardware.
- [ ] I know to check VFO vs. Memory mode first if the keypad is ignoring me.
- [ ] I know to check where the A/B active arrow is pointing before I transmit.
- [ ] I know to check my physical battery connection if the radio acts erratic or restarts.
- [ ] I am ready to use the symptom chapters ahead to fix my specific problem.

## If you're stuck, go to...

- *Chapter 9* (if your symptom is: you can hear them, but they can't hear you).
- *Chapter 10* (if your symptom is: the radio is completely silent or blasting static).
- *Chapter 12* (if your symptom is: your saved channels keep disappearing).

Chapter 9

# I Can Hear Them, But They Can't Hear Me

**The Symptom:** You are tuned perfectly to a local repeater. You hear the automated weather bot, or you hear two older guys talking about their antennas with crystal-clear audio. Taking a deep breath, you press the Push-To-Talk (PTT) button, proudly announce your call sign, and release the button.

Dead silence. Crickets. Nobody replies. You don't even hear the polite, robotic "beep" from the repeater tower confirming it heard your transmission.

This is undeniably the most frustrating beginner symptom in the entire hobby because it makes you feel completely invisible, or worse, like you've broken your brand new gear. But don't worry, this is a rite of passage. Almost every new radio owner experiences this exact moment. The fix is almost always a simple software setting related to the invisible "bouncer" at the repeater's door.

**What you'll accomplish:**

- Diagnose and fix incorrect Offset (Duplex) settings by visually watching the screen.
- Diagnose and fix incorrect CTCSS/Tone settings (the VIP password).
- Verify your physical power output and understand your environmental range limitations.

**What you need:**

- Your Baofeng radio.
- The original RepeaterBook listing (Frequency, Offset, and Tone) for the specific tower you are trying to hit.

## Fix Tree Step 1: Confirm the Duplex (Offset)

Remember the "two lanes of the highway" concept from Chapter 4? Repeaters listen on one lane and talk on another to avoid a deafening feedback loop. If you don't automatically jump over to the repeater's listening lane the exact moment you press PTT, you are just shouting at a brick wall.

1. Go to the channel or frequency you are having trouble with.
2. Press the **PTT** button briefly while keeping your eyes glued to your Baofeng's screen.
3. *Did the frequency numbers physically change while you were holding the button down?*

- **If NO:** Your offset is missing. The numbers stayed stubbornly frozen, meaning you are trying to talk to the tower's speaker instead of its microphone. You need to go into **Menu 25 (SFT-D)** and set it to + or - according to your notes. Then, go to **Menu 26 (OFFSET)** and ensure the amount (usually 000600 or 005000) is typed in.
- **If YES:** You literally saw the numbers drop or jump (e.g., from 146.700 down to 146.100). The math is working perfectly. Move to Step 2.

## Fix Tree Step 2: Confirm the Transmit Tone

If the radio is jumping to the correct frequency lane, but the tower is still completely ignoring you, the tower's automated "bouncer" is rejecting your signal. You are standing at the right door, but you aren't whispering the secret VIP password.

1. Check your repeater notes. Does the tower require an "Uplink Tone" (like 100.0)?
2. Press **MENU** and go to **Menu 13 (T-CTCS)**.
3. Is it set to OFF, or is it set to the wrong number?
4. Unlock the menu, use the arrows to carefully select the exact tone listed in your notes, and press **MENU** to confirm. Remember, this tone is sub-audible, you will never actually hear it with your human ears, but the tower's computer will.

*STOP*

**The Receive Tone Trap:** Check **Menu 11 (R-CTCS)** right now. Is there a number in there? If so, turn it **OFF**. You should almost never have a Receive Tone turned on as a beginner. Setting a receive tone is like putting heavy earplugs in your own ears. You are telling your Baofeng, *"Ignore everyone and keep my speaker muted unless they transmit a secret password to me."* Because repeaters rarely send passwords back to you, your radio will stay totally silent, even when people are trying to answer your call.

## Fix Tree Step 3: Check Your Power Level

If your offset is right and your tone is perfect, but the tower still can't hear you, you might just be whispering too quietly for the distant microphone to pick you up.

Think of transmit power like the volume of your own voice. Your Baofeng can transmit at High Power (usually 5 to 8 watts) or Low Power (1 watt). Low Power is like an indoor conversational voice, great for talking to a buddy in the same campsite and saving battery life, but terrible for shouting across town.

1. Look at your screen. Do you see a tiny **"L"** on the left side? If so, you are in Low Power mode.
2. *The Fix:* Tap the **# / Padlock key** once quickly (do not hold it down). The "L" should immediately disappear, putting you back into High Power mode (the equivalent of shouting). Try your transmission again.

## Fix Tree Step 4: Move Your Body (The Line-of-Sight Rule)

VHF and UHF radio waves act exactly like a beam of light from a flashlight. They do not bend around the earth, and they absolutely hate thick concrete, steel framing, and dense rocky hills.

If a flashlight beam cannot pass through the wall of your house, neither can your radio signal. If you are sitting on the couch in your basement trying to hit a repeater 15 miles away, the laws of physics are working against you.

- *The Fix:* Change your environment. Walk outside into the yard. Stand on an elevated back deck. If you are sitting in your car, remember that the vehicle is essentially a giant metal cage that bounces your signal right back at you, roll the window down or physically step out of the vehicle. Sometimes, just holding the radio three feet higher in the air or taking ten steps to the left can be the difference between a perfectly clear signal and being completely ignored by the tower.

## Quick Win Checklist

- [ ] I watched my screen while pressing PTT to confirm my frequency is actually shifting (Offset).
- [ ] I checked Menu 13 (T-CTCS) to ensure my Transmit Tone password is set correctly.
- [ ] I checked Menu 11 (R-CTCS) to ensure I didn't accidentally lock my own radio by setting a receive tone.
- [ ] I verified my radio is in High Power mode (no tiny "L" on the screen).
- [ ] I physically moved outside or higher up to improve my line-of-sight to the tower.

## If you're stuck, go to...

- *Chapter 11: I Can Hit One Repeater but Not Another* (if you applied all these fixes, and it works for Tower A, but Tower B is still stubbornly ignoring you).

Chapter 10

# I Can't Hear Anything (Or It's Random and Noisy)

**The Symptom:** Your buddy is standing fifty feet away from you, pressing his PTT button and talking directly to you. But your radio is dead silent. Or, alternatively, your radio sits quietly on the desk and then randomly blasts you with five seconds of terrifying, ear-piercing static, then goes silent again.

If you can't hear incoming traffic, your radio is essentially a paperweight. Let's get your ears open.

**What you'll accomplish:**

- Fix the most common squelch and volume errors.
- Diagnose "Step Size" issues that push you off the correct frequency.
- Tame a radio that randomly blasts static for no reason.

**What you need:**

- Your Baofeng radio.

## Fix Tree Step 1: The Squelch Wall

The absolute most common reason a radio goes deaf is that the invisible static gate (the Squelch) has been built too high.

1. Press the **MONI** button on the side of the radio. Do you hear rushing static?
    - *If YES:* Your speaker works. The problem is your settings.
    - *If NO:* Your volume knob is turned all the way down, or your battery is dead.
2. Press **MENU** and go to **Menu 0 (SQL)**.
3. If this number is set to 5, 8, or 9, your squelch is violently high. It is blocking out legitimate signals.
4. *The Fix:* Change the Squelch down to Level **1** or **2**.

## Fix Tree Step 2: The Receive Tone Trap

We mentioned this in the last chapter, but it bears repeating because it ruins so many beginners' days.

If you set a Receive Tone (CTCSS/DCS), you are telling your Baofeng: *"Keep the speaker shut off unless the incoming signal has a secret VIP password."* If your buddy is transmitting to you on a standard simplex channel, he is not sending a password. Therefore, your radio ignores him.

- *The Fix:* Press **MENU** and go to **Menu 11 (R-CTCS)**. Ensure it is set to **OFF**. Do the same for **Menu 10 (R-DCS)**. Ensure it is also set to **OFF**. Keep your receiving doors wide open.

## Fix Tree Step 3: The "Step Size" Rounding Error

**The Symptom:** You try to type 146.520 into the keypad. But when you hit the last zero, the radio magically changes it to 146.525. It feels like the machine is mocking you.

Your radio tunes through frequencies in specific "steps," like walking up a staircase. If your step size is too large (like 25.0kHz), the radio's computer physically cannot step on 146.520. It rounds up to the nearest large step.

- *The Fix:* Press **MENU** and go to **Menu 1 (STEP)**. Change this value to **2.5K**. This is a tiny, precise staircase that will allow you to type in almost any exact frequency without the radio fighting your inputs.

## Fix Tree Step 4: Random Blasts of Static

**The Symptom:** The radio is quiet, then suddenly blasts raw static for a few seconds, then stops. Nobody is talking.

This usually means your Squelch is set *too low* (Level 0 or 1), or you are sitting near a massive source of electromagnetic interference. Cheap LED lightbulbs, computer monitors, and unshielded power supplies can actually leak radio waves that trick your Baofeng into opening its squelch gate.

- *The Fix:* Move the radio away from your computer desk or the running engine of your car. If the random static bursts continue, bump your Squelch (**Menu 0**) up from Level 1 to Level 2 or 3.

## Quick Win Checklist

- [ ] I used the physical MONI button to verify my speaker actually produces sound.
- [ ] I checked Menu 0 (SQL) and ensured my squelch is reasonably low (Level 1, 2, or 3).
- [ ] I checked Menu 11 (R-CTCS) and Menu 10 (R-DCS) and made sure they are turned completely OFF.
- [ ] I set my Step Size (Menu 1) to 2.5K so the radio stops auto-changing my frequencies.
- [ ] I moved my radio away from computer monitors and LED lights to prevent random static bursts.

## If you're stuck, go to...

- *Chapter 12: My Channels Won't Save* (if you applied these fixes in VFO mode, but the radio goes deaf again as soon as you switch to Memory Mode).

# Narrative Interlude: When the Radio Goes Silent

## The Silent Morning

**Character:** Mark Jensen (Midwest Suburbs)

**Skills reinforced:**

- Triage flow
- Mode/Frequency/Tone checkpoints
- Squelch fixes
- Step-size sanity

**Where to learn the steps:** Chapters 8, 9, & 10

It was 6:15 AM on a Tuesday. Outside Mark Jensen's kitchen window, a fresh layer of Minnesota snow was furiously burying his driveway. Inside, the house was quiet, smelling of dark roast coffee and the comforting hum of the refrigerator.

Mark sat at the kitchen island, taking a sip of his coffee. This was his moment. He had his Baofeng UV-5R sitting proudly on the granite countertop. Over the weekend, he had successfully programmed his local repeaters, talked to Ron in the driveway, and felt like a certified communications expert. He knew there was a "morning commute net" on the Twin Cities repeater every weekday at 6:30 AM, where local hams swapped weather and traffic reports.

He reached out and confidently twisted the volume knob clockwise.

*"BEEP. FREQUENCY MODE."*

Mark waited for the familiar crackle of morning chatter.

Nothing.

Dead silence.

He frowned. He picked up the radio and held it to his ear like a seashell, hoping to catch even the faintest hiss of static. Nothing. The green LCD screen stared blankly back at him, displaying 146.700.

"Okay," Mark whispered. "Maybe nobody is awake yet. Let's check the weather."

He pressed the orange **VFO/MR** button to ensure he was in the manual typing mode. He started to punch in the National

Weather Service frequency he knew by heart: 1 6 2 5 5 0.

But as his finger hit the final zero, the radio gave a sharp, angry beep. The screen physically deleted his zero and replaced it with a five. It now read 162.555.

"What?" Mark muttered. He cleared it and tried again. 1 6 2 5 5 0.

*Beep.* The screen aggressively jumped to 162.555 again.

Mark's stomach dropped. The radio wasn't just deaf; it was actively fighting his inputs. He tried to type in a different local repeater, 146.520. The radio hijacked his typing and forced it to 146.525.

"It's broken," Mark said aloud, the panic rising in his chest. "I left it on the charger too long. Or the cold air by the window fried the motherboard. It's glitching."

His mind immediately went to the worst-case scenario. The cheap Chinese electronics had failed. His brief, glorious weekend as a radio operator was over. In a desperate bid to fix it, Mark started button-mashing. He pressed the UP arrow. The numbers went wild. He pressed the blue A/B button frantically. He hit MENU and started scrolling aimlessly through the 40 different options, looking for a magic "Fix Everything" button.

He landed on Menu 40: RESET.

He pressed MENU to unlock it. He scrolled to ALL.

He pressed MENU again. The radio flashed SOURCE?

Mark's thumb hovered over the MENU key. One more press, and he would nuke the radio from orbit. He would wipe out all his saved channels, his perfectly calculated offsets, and his transmit tones. He would be back to square one, sitting in the kitchen with a beeping brick.

"Having fun with the tactical walkie-talkie at six in the morning?"

Mark jumped. His wife, Jen, was standing in the kitchen doorway, wrapped in a heavy fleece robe.

"It's dead, Jen," Mark said miserably, pulling his thumb away from the nuclear reset button. "It won't let me type the right numbers, and the speaker is completely blown. I'm going to have to factory reset the whole thing."

Jen walked over to the coffee maker, poured herself a mug, and glanced at the yellow *Ghostwriter Guide* sitting on the counter next to Mark.

"Did you do the triage thing?" she asked, gesturing vaguely with her mug. "The book says not to panic. It says to check the 'Big Three' first."

Mark sighed. "Jen, it's fundamentally broken. The math chip is scrambled."

"Humor me, Mission Control," she said, leaning against the counter.

Mark pressed EXIT to escape the dangerous reset menu. He opened the guide to Chapter 8: *How to Diagnose in 60 Seconds.* He read the first rule. *Shift your mindset from "my radio is permanently broken" to "my radio just has a setting out of place."*

"Okay," Mark said, taking a calming breath. "Checkpoint one. Mode. Am I in VFO mode?" He looked at the screen. There were no tiny channel numbers on the right. "Yes. Mode is correct."

"Checkpoint two," Jen read over his shoulder. "Squelch and Monitor. See if the speaker actually works."

Mark pressed the small **MONI** button on the side of the radio.

*SHHHHHHHHHHHHHHHHHH!*

The rushing wall of white noise blasted out of the speaker, loud and clear. Mark almost dropped the radio.

"Okay!" Mark gasped, his eyes widening. "The speaker isn't blown. It works perfectly."

He let go of the button, and the radio went dead silent again. He flipped to Chapter 10. *If the radio is dead silent but the MONI button works, your squelch is too high.* "Wait a minute," Mark said, his brain finally engaging. He pressed MENU and typed 0.

The screen displayed SQL. Mark pressed MENU to unlock it. The squelch was set to 9. The absolute maximum setting. A literal brick wall.

"Why is this on 9?" Mark asked. Then he remembered. Yesterday afternoon, he had let Owen hold the radio to show him how bright the built-in LED flashlight was. Owen's teenage thumbs had been flying all over the keypad like he was playing a video game.

"Owen," Mark whispered, shaking his head. He changed the squelch back down to 2 and saved it.

"Alright, the gate is open," Jen said. "Now why wouldn't it let you type the numbers?"

Mark flipped the page to the next fix. *The Step-Size Rounding Error.* "The staircase," Mark read aloud. "If the radio won't land on a specific number and keeps rounding up, the step size is too large."

He pressed MENU and typed 1. The screen displayed STEP. He unlocked it. Sure enough, the step size was set to a massive 25.0K. Owen had scrambled that, too. Mark smiled, feeling the anxiety completely melt away. He changed the step size down to a precise 2.5K and pressed MENU to save it. He exited back to the main screen. With a newfound swagger, he typed 1 6 2 5 5 0.

The radio accepted every single digit flawlessly. The squelch gate, now properly lowered to Level 2, instantly opened.

*"...winter storm warning remains in effect until noon today. Expect heavy snowfall and hazardous driving conditions..."* the robotic weather voice droned, loud and clear.

Mark set the radio down on the granite counter and took a long, victorious sip of his coffee. He looked at Jen. "Fixed it," Mark said, tapping the side of his head. "Just took a little advanced technical diagnostics."

"Right," Jen smiled, taking her coffee toward the living room. "Next time, maybe check the settings before you initiate the self-destruct sequence."

Mark switched the radio over to Memory Mode and dialed up Channel 1. Right on cue, the morning commute net crackled to life, filled with local hams complaining about the snow. Mark listened, completely relaxed. He hadn't needed to wipe his radio. He hadn't needed to buy a new one. He just needed to stop, breathe, and trust the troubleshooting tree. He had successfully diagnosed and cured a seemingly dead radio in under a minute.

"Checkpoints beat panic," Mark said to himself, watching the green screen glow.

## Takeaway

When the radio fails or acts erratic, slow down. Do not immediately resort to a factory reset. The vast majority of "broken" radios simply have a setting out of place (like Squelch set to 9, or Step Size set too high). Use the 60-second triage flow to methodically check your Mode, your Speaker (MONI button), your Squelch, and your Step Size.

**Bridge:** Okay, Mark has his local settings dialed in perfectly. But why does one repeater work flawlessly, while another one completely ignores him?

Chapter 11

# I Can Hit One Repeater but Not Another

**The Symptom:** You followed the instructions in Chapter 5 perfectly. You programmed "Repeater Alpha" into Channel 1, and it works flawlessly. You get signal reports, and everyone hears you. Confident, you use the exact same process to program "Repeater Bravo" into Channel 2. But when you switch to Channel 2 and press PTT... nothing. The tower ignores you.

You know your radio works, and you know you understand the menus. So why is one tower accepting you and the other shutting you out?

**What you'll accomplish:**

- Understand the difference between CTCSS and DCS tones.
- Check for "odd splits" (non-standard offset amounts).
- Learn the signs of a busy or "linked" repeater that requires different timing.

**What you need:**

- Your Baofeng radio.
- The RepeaterBook listings for *both* the working tower and the broken tower to compare them.

## Fix Tree Step 1: The DCS vs. CTCSS Trap

Not all VIP passwords are created equal.

In Chapter 4, we learned about **CTCSS** (analog tones, like 100.0). This is what 90% of repeaters use. However, some newer or private repeaters use a digital password system called **DCS** (Digital Coded Squelch).

If you look at the repeater listing for Tower Bravo and the tone says something like D023N or 023, you are dealing with a DCS tower.

If you try to type 023 into the regular CTCSS menu (Menu 13), the Baofeng won't let you, or it will transmit the wrong type of password. The analog bouncer is looking for an analog password, and the digital bouncer is looking for a digital one.

- *The Fix:* If the tower uses DCS, leave Menu 13 (T-CTCS) completely **OFF**. Instead, press **MENU** and go to **Menu 12 (T-DCS)**. Enter your digital code here.

## Fix Tree Step 2: The Non-Standard Offset (Odd Splits)

Most repeaters follow polite rules. In the 2-meter band, they almost always use a 0.600 MHz jump. In the 70-centimeter band, they almost always use a 5.000 MHz jump.

But sometimes, a local radio club has to put a tower in a crowded city where the standard frequencies are already taken. They have to use a weird, custom offset amount (like a 1.200 jump or a 7.600 jump). This is called an "Odd Split."

1. Look closely at the RepeaterBook listing for Tower Bravo.
2. Does the Offset exactly match the standard?
3. *The Fix:* If it is an odd split, you must go into **Menu 26 (OFFSET)** and manually type in that specific weird number (e.g., 001200) before you save the channel.

**[FIELD NOTE] The Save Override:** Remember, if you are fixing an existing memory channel, the Baofeng will not let you overwrite it easily. You must delete the channel (**Menu 28**) before you can re-save the new, corrected offset data into that slot (**Menu 27**).

## Fix Tree Step 3: Linked Repeaters and Timing

Sometimes the settings are 100% correct, but the tower still seems to ignore you or cuts off the first half of your sentence.

Many modern repeaters are "linked" to other towers across the state via the internet. When you press the PTT button, Tower Bravo has to wake up, send a signal across the internet to Tower Charlie, wake that tower up, and open the gates. This process takes time.

If you press PTT and instantly say "This is K1ABC," you spoke while the internet link was still booting up. The tower missed your call sign entirely.

- *The Fix:* The "Press, Pause, Speak" rule becomes critical here. Press the PTT button. Count *"One-Mississippi"* in your head. *Then* speak. Give the digital network time to open the doors for you.

## Quick Win Checklist

- [ ] I have compared the listings for Tower A and Tower B to look for differences.
- [ ] I checked if Tower B requires a digital DCS code (Menu 12) instead of a standard CTCSS tone (Menu 13).
- [ ] I checked if Tower B uses a weird, non-standard offset amount that requires me to change Menu 26.
- [ ] I am using the "Press, Pause, Speak" method to allow linked repeaters time to wake up.
- [ ] I remembered to completely delete the broken memory channel before trying to save my corrected settings into it.

## If you're stuck, go to...

- *Chapter 14: CHIRP / Cable Problems* (if you have spent an hour fighting the menus and want to use a computer to just copy-paste the correct repeater data).

Chapter 12

# My Channels Won't Save (Or They Disappear)

**The Symptom:** You painstakingly spend ten minutes programming a repeater into your radio. You carefully type the frequency, set the offset math, and dial in the correct tone password. You go to Menu 27, save it to Channel 001, and hear the robotic voice proudly confirm the save. But when you press the orange VFO/MR button to switch to Memory Mode and test your hard work, Channel 001 is completely empty. Or worse, it just shows the old, useless factory setting that was there yesterday.

You feel like the radio is intentionally gaslighting you. You might even wonder if the internal memory chip is physically defective.

Take a breath. The Baofeng memory system is not broken; it is just incredibly stubborn and highly literal. It follows a very rigid set of computer-logic rules dating back to its original engineering. If you violate even one of these invisible rules, it silently rejects your work without giving you a helpful error message or a warning beep. Let's outsmart it together.

**What you'll accomplish:**

- Understand the crucial "Sweep the Floor" rule to prevent silent overwrite failures.
- Diagnose the "Wrong Mode" error that tricks you into thinking your saves are working.
- Master the quirky "Double Save" sequence so your repeater channels actually stick permanently in the vault.

**What you need:**

- Your Baofeng radio.
- Your written notes for the specific channel you are trying to save.

## Fix Tree Step 1: Confirm You Are in VFO Mode

You physically cannot build or save a new channel while you are in Memory (Channel) Mode. Think of Memory mode like a locked glass display case, you can look at the channels, and you can use them, but you can't alter them. The radio intentionally locks the keypad from making structural changes to protect your existing emergency presets from accidental bumps. If you try to save something while in Memory mode, the radio will pretend to accept the button presses, but absolutely nothing will happen behind the scenes.

- *The Fix:* Look at the right side of your screen. Do you see tiny channel numbers (like 001 or 127)? If yes, press the orange **VFO/MR** button once until those numbers completely disappear. You must build the channel on the blank, unlocked canvas of VFO mode before you can move it into the display case.

## Fix Tree Step 2: The "Sweep the Floor" Rule

This is the number one reason channels fail to save, and it catches almost every beginner off guard. The Baofeng will not let you paint over a dirty canvas. Furthermore, it will *never* ask you, "Are you sure you want to overwrite this channel?"

If Channel 005 already has a frequency saved in it (even a random factory test frequency you didn't put there), and you try to save a brand new repeater into Channel 005, the radio's computer acts like a bouncer. It silently rejects the new data to protect the old data, leaving your new frequency lost in the void.

- *The Fix:* You must explicitly sweep the floor and delete the old channel before you save to it. Press **MENU**, type **2 8 (DEL-CH)**, unlock the menu, select your target channel number using the arrow keys, and press **MENU** to confirm the deletion. You will hear the robotic voice confirm the channel is cleared. Once the slot is truly empty, the radio will gladly accept your new data.

## Fix Tree Step 3: The Double-Save Sequence

As we learned in Chapter 5, a repeater channel actually consists of two completely separate frequencies working together: the listening lane and the talking lane. Most modern, expensive radios automatically bundle these together for you. The Baofeng does not. It requires you to save them one at a time, manually, in a very specific, quirky order.

If you only save the listening lane, the channel will appear on your screen and you will hear traffic, but it will instantly fail when you press the PTT button because the radio doesn't know where to jump.

- *The Fix:* Let's review the exact rhythm and sequence in **Menu 27 (MEM-CH)**.

   a. Once your frequency, offset, and tone are set perfectly in VFO mode, go to Menu 27.

   b. Unlock it, pick your newly emptied channel slot, and press **MENU**. You will hear the voice proudly say "Receiving Memory."

**[STOP]** Do not press Exit yet! The job is only half done. You've only packed the listening half of the suitcase.

1. Press **MENU** again to unlock the menu a second time. The little black arrow will drop down.
2. Ensure the exact same channel slot is still selected, and press **MENU** a third time. You will hear "Transmitting Memory."
3. Now you can safely press **EXIT**. The complete package, listening lane, talking lane, and tones, has been successfully bundled and stored.

*FIELD NOTE*

**Simplex Channels Only Need One Save:** If you are saving a pure radio-to-radio "simplex" channel (like your family channel from Chapter 7) that has no offsets and no tones, your listening and talking lanes are the exact same frequency. In this specific case, you only need to perform the save sequence once. The "Double Save" is exclusively required for complex repeaters.

## Quick Win Checklist

- [ ] I confirmed I am entirely in VFO mode before attempting to program any settings.
- [ ] I explicitly used Menu 28 (DEL-CH) to sweep the floor and delete my target memory slot before trying to save new data into it.
- [ ] I remember to perform the "Double Save" in Menu 27 when programming repeaters, ensuring I listen for both the "Receiving" and "Transmitting" voice prompts.
- [ ] I immediately switched back to Memory Mode (VFO/MR button) to visually test that my channel was actually stored.

## If you're stuck, go to...

- *Chapter 14: CHIRP / Cable Problems* (if manual programming is causing you too much stress and you are ready to use a computer cable to simply drag-and-drop your channels).

# Narrative Interlude: Why One Repeater Works and Another Won't

**Character:** Nate Brooks (Southeast)

**Skills reinforced:**

- CTCSS vs. DCS
- Tone-squelch pitfalls
- Deliberate channel overwrite
- Save discipline

**Where to learn the steps:** Chapters 11 & 12

The asphalt parking lot of the North Charleston Community Center was already radiating waves of intense morning heat. Nate Brooks sat in the driver's seat of his Tacoma, the air conditioning running on high, staring at a single piece of paper on his center console.

It was his CSCE, his Certificate of Successful Completion of Examination. He had just passed his Technician class amateur radio test. He was officially licensed. The government had assigned him a call sign, KN4XYZ, which meant his self-imposed ban on pressing the Push-To-Talk button was officially lifted. He was no longer just a listener; he was an operator.

But Nate wasn't the type to celebrate without finishing the job. He pulled out the bright yellow, laminated "Home Profile" index card from his tactical pouch. Channel 1, the Primary City Repeater, was locked in and working perfectly. However, for a hurricane preparedness plan to be truly viable, he needed redundancy. He needed Channel 2: The North County Backup Repeater.

He opened his notebook to the RepeaterBook printout he had generated earlier that week. *North County RPT. Frequency: 444.100. Offset: +5.000. Tone: D023N.*

"Alright," Nate muttered, slipping into his procedural mindset. "Let's build the backup route."

He picked up the Baofeng UV-5R, pressed the orange VFO/MR button to enter the manual typing mode, and keyed in 4 4 4 1 0 0. He went to Menu 25 and set the shift direction to positive. He went to Menu 26 and carefully typed 005000.

Next came the password. Nate went to Menu 13, T-CTCS, the transmit tone menu he had used for Channel 1. He scrolled through the numbers, looking for 023. It wasn't there. The lowest number was 67.0.

Frowning, Nate started pushing buttons. He went into Menu 11, R-CTCS, thinking maybe the backup tower required a receive tone instead. He selected a random tone just to see if it would force the radio to accept his input. He was guessing, and for a man who lived by checklists, guessing was a dangerous spiral.

Frustrated, he decided to just force the save and see what happened. He navigated to Menu 27, selected Channel 002, and pressed MENU. The robotic voice chimed, *"Receiving Memory."* He pressed it again. *"Transmitting Memory."*

Nate confidently pressed the VFO/MR button to drop back into his locked-down Memory mode. He clicked the UP arrow to Channel 2.

The screen read 155.500. It was a useless factory default frequency. His meticulously typed 444.100 was nowhere to be found.

"What?" Nate stared at the screen. He tried the entire process again, typing faster this time. He saved it to Channel 2. He checked Memory mode. Still 155.500.

The radio was silently rejecting his programming. Furthermore, when he clicked back down to his perfectly programmed Channel 1 to make sure the radio wasn't entirely broken, he was met with dead silence. He couldn't even hear the static tail of the primary repeater anymore.

His clean, disciplined communication plan was falling apart. He had introduced too many variables, changed too many settings without documenting them, and now he was locked out of his own system.

"Having trouble with the new ticket, KN4XYZ?"

Nate rolled his window down. Ms. Patel, the lead volunteer examiner who had just graded his test thirty minutes earlier, was walking toward her SUV. She was in her late fifties, possessed an aura of absolute calm authority, and had a massive VHF/UHF antenna mounted to the roof of her vehicle.

"I've compromised the baseline," Nate said flatly, holding up the Baofeng. "I'm trying to program the North County repeater into slot two. It refuses to save over the factory junk, and in trying to fix it, I've somehow deafened slot one."

Ms. Patel leaned against the door of his truck. "Let's look at the data. What's the RepeaterBook listing for North County?"

Nate read it off his notepad. "444.100. Plus offset. Tone is D023N."

"Stop right there," Ms. Patel said gently. "Look at that tone again. It doesn't say 100.0 or 141.3. It has a 'D' in front of it. That stands for Digital. You're trying to put an analog key into a digital lock."

Nate looked at the notebook. He had completely glossed over the letter 'D'.

"Menu 13 is for analog CTCSS tones," she explained. "When you couldn't find 023 in there, you started guessing. You went into Menu 11 and set a Receive Tone, didn't you?"

Nate checked the screen. There was a tiny CT icon illuminated on the left side of the display. "Yes. I changed a variable I didn't fully understand."

"And by doing that, you put a bouncer on your own speaker," Ms. Patel said, tapping the edge of his window frame. "Your radio is now ignoring every signal in Charleston because they don't have the secret password you just demanded. Turn Menu 11 off. Never use a receive tone unless you absolutely have to."

Nate immediately went to Menu 11 and set it to OFF. The tiny CT icon vanished.

"Now, for the memory failure," she continued. "The Baofeng is a stubborn machine, Nate. It will never let you paint over a dirty canvas. If Channel 2 already has 155.500 saved in it, the radio will silently reject your new save to protect the old data. You have to sweep the floor first."

"Deliberate overwrite," Nate realized, a wave of clarity washing over him. "I have to delete it before I can replace it."

"Exactly. Listen first. Verify receive. Then change one thing at a time," Ms. Patel said, giving him a warm smile. "Welcome to the airwaves, Nate. You'll get it." She turned and walked toward her SUV.

Nate took a deep breath. He looked at his radio and his notebook. Procedure. He could do procedure.

He went to **Menu 28 (DEL-CH)**. He selected Channel 002 and pressed MENU. *"Confirm,"* the radio said. The slot was now officially empty.

He went to VFO mode. He typed 4 4 4 1 0 0. Offset was already set to + and 005000.

Now for the password. He ignored Menu 13 entirely. He went to **Menu 12 (T-DCS)**. He scrolled through the digital codes until he found D023N. He pressed MENU to confirm it.

Finally, save discipline. He went to **Menu 27 (MEM-CH)**. He selected his freshly swept Channel 002. Press MENU. *"Receiving Memory."* Press MENU again. Arrow drops. Press MENU a third time. *"Transmitting Memory."*

Nate pressed the VFO/MR button to switch to his saved channels. He clicked up to Channel 2. The screen proudly displayed 444.100. It had saved perfectly.

He held the radio up to his mouth. It was time for a true, legal confidence check. He pressed the PTT button, watched the numbers automatically jump up by 5.000 to the transmit lane, and spoke clearly.

"KN4XYZ, testing."

He released the button. The numbers jumped back. A split-second later, a crisp burst of static erupted from the speaker, followed by a polite *beep* from the distant tower.

*"Loud and clear, KN4XYZ. Good morning,"* an older gentleman's voice replied.

Nate smiled. He reached for his black permanent marker and placed a confident checkmark next to Channel 2 on his yellow index card. His redundancy was secured. He had two working repeaters, programmed correctly, saving reliably, and his comms plan was completely restored.

## Takeaway

A good plan is small and clean. Random tweaks are how plans die. If a repeater listing uses a "D" before the tone (like D023N), it is a Digital Code. Leave the standard CTCSS menu alone and use Menu 12 (T-DCS) instead. Above all, the Baofeng will silently reject your programming if you try to save over an existing channel, you must explicitly use Menu 28 to delete the channel slot before attempting to save new data into it.

**Bridge:** If I'm going to do this, I need my programming and backups to survive reality.

## Chapter 13

# Scanning Doesn't Work (Or It Skips Everything)

**The Symptom:** You want to hear what is happening in your city during an event or emergency. You press and hold the Scan button (* / SCAN), hoping the radio will act like a high-end police scanner and automatically find local traffic for you to listen to. Instead, the frequency numbers blur past at lightning speed and never stop. Or, worse, it immediately stops on a channel that is just blasting pure, deafening static, and it gets stuck there forever, forcing you to turn the radio off in frustration.

Scanning on a Baofeng is famously quirky. Your device is a two-way radio first and a scanner second. It is not as smart, as fast, or as highly filterable as a dedicated $400 digital police scanner. But once you understand how its simple, mechanical "brain" actually works, you can make it a highly effective tool for gathering local intelligence and monitoring your neighborhood.

**What you'll accomplish:**

- Understand why your squelch setting is completely breaking your scan.
- Define and set the correct "Scan Resume" behavior in plain English so you stop missing the ends of conversations.
- Learn how to deal with "nuisance channels" that constantly hijack your scanner and interrupt you.

**What you need:**

- Your Baofeng radio.

## Fix Tree Step 1: The Squelch Block

The Baofeng's scanning feature doesn't have a sophisticated computer brain; it relies entirely on the invisible Squelch gate that we set up way back in Chapter 2.

When the radio is in scan mode, it is rapidly flipping through frequencies. It only knows to stop when it physically collides with a signal strong enough to break through the squelch gate.

- **[STOP] Never scan on Level 0.** The squelch gate is completely removed. The radio will immediately "detect" the normal background static on the very first channel it checks. It will stop scanning instantly, thinking it found a real signal, and just blast static at you forever.
- If your Squelch is set to **Level 9**, the gate is a massive, impenetrable wall. The radio will scan right past legitimate, active human conversations because the signals aren't strong enough to break over that incredibly high wall. You will watch the scanner run in circles for hours and hear absolutely nothing.

- *The Fix:* Before you ever press the scan button, ensure **Menu 0 (SQL)** is set to **Level 2 or 3**. This provides the perfect Goldilocks balance, it allows the scanner to confidently skip over empty static, but immediately hits the brakes when a real human voice appears.

## Fix Tree Step 2: The "Scan Resume" Mystery

**The Symptom:** Your scanner is working. It successfully finds a local repeater where two people are talking about a road closure. It stops so you can listen. But five seconds later, right in the middle of a critical sentence, the radio brutally cuts them off, beeps, and starts scanning again! It feels like someone grabbing the TV remote and changing the channel right at the climax of a movie.

This happens because the Baofeng has three entirely different "Scan Resume" modes, and the factory default is arguably the most annoying one for a beginner.

1. Press **MENU** and go to **Menu 18 (SC-REV)**. This stands for Scanner Resume.
2. Unlock it. You will see three obscure abbreviations: **TO**, **CO**, and **SE**. Here is what they actually mean in plain English:

- **TO (Time Operation):** The radio stops on an active signal, starts a strict 5-second countdown timer, and then blindly resumes scanning, *even if the person is still actively talking*. This is the annoying factory default. It is meant for just sampling the airwaves, not for actually listening to a conversation.
- **CO (Carrier Operation):** The radio stops on an active signal and patiently stays there for as long as the person is holding down their microphone. The moment they un-key their radio and the channel goes quiet, your Baofeng waits 3 seconds to see if anyone replies, and then politely resumes scanning.
- **SE (Search Operation):** The radio stops on the very first active signal it finds and stays there forever. The scanning function is permanently canceled the moment it finds a single voice.
- *The Fix:* Change Menu 18 to **CO (Carrier Operation)**. This is the absolute gold standard for emergency scanning. It lets you listen to the entire conversation naturally, and automatically goes back to hunting for new signals only when the conversation is actually over.

## Fix Tree Step 3: Nuisance Channels

Sometimes, you have a local NOAA weather broadcast or a noisy digital dispatch frequency saved in your memory bank. Because weather stations broadcast continuously without ever taking a breath, and digital data streams sound like an angry 1990s dial-up modem playing on repeat, they will hold your scanner hostage. Every single time you run a scan, the radio stops on this noisy channel and refuses to leave.

- *The Fix:* You ultimately need to lock that specific channel out of the scan list. Unfortunately, you cannot easily do this from the radio's physical keypad. You can only skip these nuisance channels permanently by using a computer programming cable and the CHIRP software, which we will cover in the very next chapter.

*GOOD ENOUGH*

If you don't have a computer cable yet, or if you are out in the woods, there is a manual, split-second workaround. When the scanner inevitably stops and gets stuck on that noisy weather channel, just reach down and tap the **Up Arrow** key once while it is stuck. This manually forces the scanner to abandon that channel and immediately resume its hunt for the next signal. You might have to do this every time it loops around, but it keeps the scanner moving.

## Quick Win Checklist

- [ ] I confirmed my Squelch (Menu 0) is set to Level 2 or 3 before initiating any scan.
- [ ] I understand how to start a scan by pressing and holding the * / SCAN key for two seconds.
- [ ] I have changed my Scan Resume setting (Menu 18) to **CO (Carrier Operation)** so the radio doesn't cut people off mid-sentence.
- [ ] I know to quickly tap the Up Arrow key to manually force the scanner past a stuck or exceptionally noisy channel.

## If you're stuck, go to...

- *Chapter 10: I Can't Hear Anything* (if your scanner clearly stops on an active channel, but you hear absolutely no audio coming out of the physical speaker).

Chapter 14

# CHIRP / Cable Problems (Optional but Common)

**The Symptom:** Programming your Baofeng by hand is a valuable rite of passage, but eventually, you will want to upgrade your workflow. You finally decided to buy the $20 USB programming cable. You eagerly want to use the free CHIRP software on your computer to type in custom channel names, import entire lists from RepeaterBook in seconds, and bypass the clumsy radio keypad entirely.

You plug the cable into your laptop, attach it to the radio, click the software, and instead of a seamless transfer, you get a terrifying red error box: **"Radio did not respond."** Instead of saving time and sanity, you are now locked in a frustrating battle with your computer's device manager.

**What you'll accomplish:**

- Understand the absolute golden CHIRP workflow (Download -> Edit -> Upload) to prevent data corruption.
- Perform a comprehensive "Sanity Check" on your physical cable connection.
- Diagnose and fix the dreaded "COM Port" error that plagues PC users.

**What you need:**

- Your Baofeng radio (fully charged).
- A Baofeng programming cable (the "BTECH PC03" FTDI cable is highly recommended over cheap knockoffs).
- A PC or Mac with the free CHIRP software installed.

## Fix Tree Step 1: The "Radio Did Not Respond" Physical Check

The Baofeng has a notoriously tight, dual-pin accessory port. It requires the programming cable to be seated incredibly deeply into the side of the radio. If it is off by even a fraction of a millimeter, or if the rubber weather flap gets pinched underneath the plastic header, the metal data contacts won't touch. Your computer will stubbornly claim the radio doesn't exist.

1. Unplug the cable from the side of the radio entirely. Check to ensure the rubber port cover is pulled completely out of the way.
2. Push the two pins back in, and apply a surprising amount of firm, even pressure with your thumb. You want the hard plastic of the cable head completely flush against the plastic casing of the radio.

3. Ensure the radio is powered **ON**, and the volume knob is turned to maximum. (This sounds weird, but the programming data actually travels through the physical speaker/microphone circuitry inside the radio as audio tones. The volume knob literally controls the data flow!)

4. Ensure the radio is tuned to a "dead" or quiet frequency with no static, active talking, or weather broadcasts. If the radio receives a radio signal and breaks squelch while the computer is trying to talk to it, the data transfer will instantly crash and fail.

## Fix Tree Step 2: The COM Port Sanity Check

If the cable is firmly seated but CHIRP still gives you a stubborn error, you are likely playing "COM Port Roulette." Your computer physically sees the USB cable, but CHIRP doesn't know which specific digital port the cable is assigned to.

This is highly common with cheap, knockoff programming cables that use counterfeit "Prolific" microchips, which Windows actively tries to block. (This is why we highly recommend spending the extra five dollars for an authentic "FTDI" cable, they are true plug-and-play).

1. In CHIRP, click **Radio** > **Download from Radio**.

2. A small dialog box will pop up. Look closely at the **Port** dropdown menu at the top.

3. *Windows Users:* You should see a list of ports named something like COM3, COM4, or COM5. If the box is completely blank, or only shows COM1 (which is usually a reserved default system port), your computer hasn't installed the digital driver for the cable.

4. *Mac Users:* Look for a uniquely named port like /dev/cu.usbserial or /dev/tty.usbserial.

5. *The Fix:* Try selecting a different COM port from the dropdown menu and clicking OK. If it fails, try the next one on the list. If no other ports exist, completely unplug the USB cable, plug it into a different physical USB port on the opposite side of your computer, wait 30 seconds for the computer to recognize the new hardware, and try again.

## Fix Tree Step 3: The Golden Workflow (Download First!)

The absolute biggest mistake beginners make with CHIRP is opening the software, creating a brand new, blank spreadsheet, typing in all their favorite frequencies, and trying to force-upload that blank document straight into the radio.

The radio will violently reject this. Every Baofeng has invisible, factory-specific firmware headers that must be present in the data file. If you force a blank spreadsheet into the radio, you overwrite the radio's "brain" with blank space, which can render the radio useless.

You must memorize and follow the Golden Workflow: **Download, Edit, Upload.**

1. **Download:** Always start your session by clicking *Download from Radio*. This pulls a safe "clone" of the radio's unique invisible system settings, along with its current channels, into CHIRP.

2. **Edit:** You will now see a spreadsheet of your radio's current memory on your computer screen. Make all of your changes safely here. Type in your frequencies, add convenient names (like "CITYRPT" or "FAMILY1", note that the Baofeng screen can only display up to 7 alphanumeric characters), and delete all the random factory junk.

3. **Upload:** Click *Upload to Radio*. Because you started with the radio's own native system file, it will gladly accept the beautifully edited version back.

*STOP*

**Never unplug the cable during a transfer.** When the green progress bar is moving across your computer screen, the radio's top LED light will flash rapidly to indicate active data flow. Do not touch the cable, do not bump the desk, and do not turn the radio off until that progress bar fully reaches 100%. Interrupting a data transfer mid-stream can "brick" the radio, scrambling its internal memory beyond repair.

**Fix Tree Step 4: The Ultimate Safety Net (Save an Image)**

Before you start getting creative, changing dozens of obscure menu settings in CHIRP, or importing 500 frequencies you don't actually need, you need to protect yourself.

As soon as you successfully *Download from Radio* for the very first time, go immediately to **File** > **Save As** in the CHIRP menu. Save this file directly to your computer's desktop as Baofeng_Factory_Backup.img.

This file is a digital time machine. It is an exact, bit-for-bit clone of your radio the day you bought it. If you ever severely mess up the internal settings, delete something critically important, or accidentally ruin your screen colors so badly you can't read the display, you can simply open this .img file in CHIRP and upload it to the radio to instantly rewind time and restore the radio to perfect factory-new condition.

## Quick Win Checklist

- [ ] I physically pushed the programming cable into the radio until the hard plastic was completely flush against the radio body.
- [ ] I turned the radio ON and turned the volume knob to maximum before attempting a transfer.
- [ ] I selected the correct COM Port or USB Serial port in the CHIRP dropdown menu (avoiding COM1).
- [ ] I memorized the Golden Workflow: Download first, Edit the spreadsheet, then Upload back to the radio.
- [ ] I created a permanent .img backup file of my factory settings acting as my digital time machine.

## If you're stuck, go to...

- *Chapter 15: Reset and Recover* (if you attempted a CHIRP upload, something went horribly wrong, and the radio screen is now blank, entirely unresponsive, or speaking in a different language).

Chapter 15

# Reset and Recover (When You've Made a Mess)

**THE SYMPTOM:** YOU'VE SPENT the last two hours falling down a rabbit hole of confusing YouTube tutorials that used a slightly different model of radio. You pushed a sequence of buttons you didn't quite understand, and now your Baofeng is acting erratic and possessed. The screen is showing weird symbols you've never seen, it stubbornly refuses to transmit, the robotic voice prompts are suddenly speaking to you in Chinese, and absolutely none of the previous troubleshooting steps are working.

You feel a sinking wave of panic, thinking you have permanently destroyed your brand new radio and turned it into an expensive paperweight.

Take a deep breath and take your hands off the keypad. You haven't broken anything. The beauty of these inexpensive, utilitarian radios is that they are practically indestructible on a software level. There is no line of code you can type that will physically break the circuit board. Sometimes, the fastest and least stressful way to fix a deeply tangled, confusing mess is to simply stop fighting it, nuke it from orbit, wipe the slate completely clean, and start fresh.

**What you'll accomplish:**

- Understand the critical difference between a "VFO Reset" and an "ALL Reset."
- Execute a factory reset safely to clear out digital gremlins.
- Fix the default language settings that revert during a reset.
- Use a structured "Rebuild Checklist" to get your Home Profile back online in under 10 minutes.

**What you need:**

- Your Baofeng radio.
- Your physical paper backup card (from Chapter 7) detailing your Top 5 Channels.

## Fix Tree Step 1: VFO Reset vs. ALL Reset

The Baofeng has two entirely different levels of amnesia you can trigger. It's incredibly important to choose the right one for your specific problem so you don't accidentally delete work you want to keep.

1. Press **MENU** and go to **Menu 40 (RESET)**.
2. Unlock the menu. You will see two options: **VFO** and **ALL**.

- **VFO Reset (The Soft Reset):** This resets all your basic menu settings (like your Squelch level, your Step Size, your display backlight colors, and battery saver timers) back to their original factory defaults. *However, it preserves your saved Memory Channels.* If your radio is acting a little weird or the screen looks wrong, but you definitely don't want to lose the 5 repeater channels you spent hours painstakingly programming, try this soft reset first.
- **ALL Reset (The Nuclear Option):** This deletes absolutely everything without mercy. It resets the menus, violently wipes out every single saved Memory Channel, and returns the entire radio to the exact pristine state it was in when it left the assembly line in China.

## Fix Tree Step 2: Executing the Nuclear Option

If the radio is severely locked up, if you accidentally turned on a strange digital signaling feature you can't turn off, or if you bought the radio used and it is clogged full of the previous owner's random junk channels, you want the ALL Reset.

1. Ensure you have your physical index card from Chapter 7 with your frequencies written down directly in front of you. (Once you press this final button, those channels are gone forever).
2. Go to **Menu 40 (RESET)**.
3. Select **ALL** and press **MENU**.
4. The screen will say SOURCE? (This is the radio's poorly translated way of asking, "Are you absolutely sure you want to do this?")
5. Press **MENU** one more time to confirm your decision. **[STOP]** This is the point of no return. Once pressed, your memory channels are permanently gone.
6. The radio will display WAIT..., physically power itself off, and turn back on. **Do not remove the battery or turn the volume knob off during this reboot.** The screen will likely be in VFO mode, displaying factory test frequencies (like 155.500 or 136.000).

*FIELD NOTE*

**Fixing the Language:** When you perform a full factory reset, the Baofeng often defaults its robotic voice prompts back to Chinese. Don't panic. To switch it back to English, press **MENU**, type **1 4 (VOICE)**, unlock the menu, use the arrows to select **ENG**, and press **MENU** to confirm.

*YOU'RE WINNING*

The ghost in the machine is officially gone. All of your accidental menu mistakes, buried sub-tones, and confusing offsets have been wiped away. You now have a perfectly clean, predictable canvas to work with.

## Fix Tree Step 3: The 10-Minute Rebuild

Now that the radio is thoroughly wiped, we need to get it combat-ready again. Because you practiced these exact skills in Part 1, what felt like an agonizing hour on your first day will now take you less than 10 minutes. You are going to build muscle memory right now.

Grab your paper backup card and follow this exact, ruthless order of operations:

**Minute 1: The Hardware Baseline** We need to lay the foundation so the radio behaves politely.

- Go to Menu 0 (SQL) and set your Squelch to 2 (so you aren't blasted by static).
- Go to Menu 1 (STEP) and set your step size to 2.5K (so you can type precise numbers).
- Go to Menu 2 (TXP) and ensure your transmit power is on HIGH.

**Minute 2-5: Rebuild the City Lighthouses** We need to restore your long-range repeater connections.

- Ensure you are in VFO Mode (no tiny channel numbers on the right). If you see channel numbers, press the orange **VFO/MR** button once to clear them.
- Program Channel 1 (Your Primary Repeater): Type the frequency, set your Offset Amount (Menu 26), set the Duplex Direction (Menu 25), and set your Transmit Tone (Menu 13). Use the careful "Double Save" sequence in Menu 27 to lock it in.
- Program Channel 2 (Backup Repeater) using the exact same methodical process.

**Minute 6-8: Add the Neighborhood Simplex Lanes** We need to restore your grid-down, direct-to-family channels.

- Program Channel 3 (Simplex 1): Type the frequency. Ensure Offset is OFF. Ensure Tone is OFF. Save it to Menu 27 using just a single save.
- Program Channel 4 (Simplex 2) identically.

**Minute 9-10: The Final Vault Check**

- Press the orange VFO/MR button to switch confidently back into Memory Mode.
- Use the arrow keys to flip through Channels 1, 2, 3, and 4. Visually verify they are all present and correct.
- Lock the keypad by holding the # key for two seconds. You will see a tiny padlock icon appear at the top of the screen, ensuring you don't accidentally ruin your fresh, beautiful rebuild.

## The True Value of Backups

You just experienced firsthand exactly why we insisted on writing down the "Home Profile" on a physical index card in Chapter 7.

Under extreme stress, human memory is notoriously terrible. Adrenaline destroys your ability to recall fine details. You will absolutely not remember if your local repeater's tone was 100.0 or 110.9 when severe weather is rolling in or trees are coming down outside. If you have to spend an hour on your smartphone trying to look up RepeaterBook data while your neighborhood is out of power, your radio is a useless brick.

- **Level 1 Backup:** The waterproof, tape-laminated index card physically taped to the back of the radio.
- **Level 2 Backup:** Take a clear photo of that index card with your smartphone and "favorite" it in your photo album so it is always accessible, even without cell service.
- **Level 3 Backup:** The CHIRP .img file securely saved to your computer's desktop (covered in Chapter 14) for instant digital recovery.

## Quick Win Checklist

- [ ] I understand the difference between a soft VFO Reset (saves my channels) and a hard ALL Reset (wipes absolutely everything).
- [ ] I have verified I possess a physical paper backup of my frequencies before triggering an ALL Reset.
- [ ] I successfully executed an ALL Reset to clear out corrupted, confusing settings.
- [ ] I used Menu 14 to change the robotic voice prompt back to English.
- [ ] I successfully used the 10-Minute Rebuild flow to restore my Squelch, Step Size, and Top 5 critical channels.
- [ ] I feel highly confident that I can completely "un-break" this radio anytime it starts acting erratic.

## If you're stuck, go to...

- *Part 3: The Family Comms Plan* (Now that your radio is perfectly tuned, clean, and backed up, you have graduated from technical troubleshooting. You are ready to start running real-world communication drills with your family!)

# Narrative Interlude: Cable Night at the Kitchen Table

**Character:** Mark Jensen (Midwest Suburbs)

**Skills reinforced:**

- Scan basics
- CHIRP workflow
- Cable/COM sanity
- Image backups
- Reset/rebuild

**Where to learn the steps:** Chapters 13, 14, & 15

"I'll just use CHIRP, how hard can it be?"

Those were the famous last words of Mark Jensen, spoken at 8:45 PM on a Thursday at his kitchen table.

Spread out before him was his laptop, a newly delivered twenty-dollar USB programming cable, and his trusty Baofeng UV-5R. Mark had successfully programmed his local repeaters by hand over the weekend, but his thumb was getting tired of the keypad. Plus, his scanner was misbehaving. Every time he tried to scan the local airwaves, it would immediately get stuck on the National Weather Service frequency and refuse to move, blasting automated forecasts about barometric pressure until he manually turned it off.

He knew CHIRP, a free, open-source radio programming software, could fix his scanner issues and let him name his channels. He had watched a five-minute YouTube video. He felt invincible.

He plugged the USB cable into his laptop, jammed the two-pronged end into the side of the Baofeng, and opened the software. A blank spreadsheet appeared. Mark spent the next twenty minutes meticulously typing in twenty different frequencies. He added local repeaters, simplex channels, and even a few fire dispatch frequencies he found online. He painstakingly typed in the offsets and the tones.

Finally, his digital masterpiece was complete. He clicked the button labeled *Upload to Radio.*

A little box popped up, loading for three agonizing seconds before flashing a harsh, red error message: **"Radio did not respond."**

"Excuse me?" Mark said to the screen. He clicked it again.

**"Radio did not respond."**

"I am right here," Mark muttered, tapping the radio. "Respond."

Assuming the USB port was acting up, Mark unplugged the cable. He moved it to the USB port on the other side of his laptop. He clicked Upload. *Error.* He unplugged it again, blew into the USB connector like an old Nintendo game cartridge, and plugged it into a dongle attached to the back of the computer. It felt like a modern technological rain dance.

*Error. Radio did not respond.*

"Is the tactical walkie-talkie rejecting your friend request?" Jen asked, walking into the kitchen to grab a glass of water.

"It's a COM port issue, Jen," Mark said defensively, rubbing his temples. "Or a driver conflict. The computer is speaking English and the radio is expecting Chinese. They are failing to negotiate terms."

"Did you read the guide?" she asked, pointing to the yellow field manual sitting under his coffee mug.

"I don't need the guide for a USB cable. You plug it in, it works. That's how computers have worked since 1998."

Jen just raised her eyebrows, took a sip of water, and walked back to the living room.

Mark sighed. He pulled the manual out from under his mug and flipped to Chapter 14: *CHIRP / Cable Problems.* He read the first section.

*The Baofeng has a notoriously tight accessory port. If the plastic is not pushed entirely flush against the radio body, the data pins will not connect.*

Mark looked at the side of his radio. The two pins were in, but there was a tiny, one-millimeter gap between the hard plastic of the cable and the radio casing. The rubber flap was slightly pinched. He pulled the cable out, moved the flap completely out of the way, and shoved the cable back in with a firm, aggressive push from his thumb. It seated with a dull *thud*. Flush.

Next, he read the volume rule. *Turn the radio on and turn the volume knob to maximum. The data travels through the audio circuits.* Mark twisted the knob all the way up.

He looked back at the book. *The Golden Workflow: Download first! Never force a blank spreadsheet into a new radio.*

Mark's stomach dropped. He had been trying to jam a blank, generic spreadsheet into the radio's brain. If the cable connection hadn't failed, he might have accidentally wiped the radio's internal firmware.

"Okay. Follow the flow," Mark whispered to himself.

He closed his masterpiece spreadsheet. He clicked *Radio > Download from Radio.*

A prompt asked him to select his COM Port. He looked at the dropdown list. It was currently set to COM1. The book explicitly warned that COM1 was usually a phantom system default. Mark clicked the dropdown and selected COM4. He selected his radio model (Baofeng UV-5R) and clicked OK.

A green progress bar appeared on the laptop screen. At the exact same time, the little LED light on the top of the Baofeng started flashing furiously. Data was flowing.

"We have a handshake," Mark grinned.

When the download finished, the screen populated with the radio's actual internal memory. There were the 5 channels from his Home Profile, exactly as he had programmed them by hand days ago.

Before he did anything else, Mark followed the ultimate safety rule in Chapter 14. He went to *File > Save As* and saved the file to his desktop as Baofeng_Factory_Backup.img.

"Time machine activated," he said. If he ever truly broke the radio in the future, he could upload that exact file and magically restore it to this moment. The fear of permanent failure completely evaporated.

Now, he could fix his scanner. He looked at the row for Channel 5, the automated National Weather Service frequency. He scrolled to the far right of the spreadsheet to a column labeled *Skip*. He clicked the box to change it to "S".

By doing that, he was telling the radio: *When I press the scan button, skip right over this noisy channel.*

He quickly typed in a few names for his channels, CITYRPT, FAMILY1, and then initiated the final step of the golden workflow. He clicked *Upload to Radio*.

The green progress bar sailed across the screen. The radio flashed, beeped loudly, and rebooted itself.

Mark unplugged the cable. The screen now proudly displayed alphanumeric names instead of just raw numbers. He pressed and held the * / SCAN key. The radio immediately began scanning through his memory channels, skipping perfectly over the annoying weather broadcast and hunting for real human voices.

It worked. He hadn't needed to try a fourth USB port. He just needed to follow the procedure: Seat the cable, pick the right COM port, download first, and save a backup.

He leaned back in his kitchen chair, extremely satisfied. He had a fully backed-up Home Profile saved to his hard drive. He had a working scanner. He had complete control over the machine.

"Cool," Mark said to the quiet kitchen, tapping his finger against the radio's plastic casing. "Now, can other humans in my house actually use this?"

## Takeaway

Backups turn frustration into a speed bump. Never try to force a blank spreadsheet into a radio. Always push the physical cable in firmly, turn the volume up, and follow the Golden Workflow: **Download from Radio -> Edit -> Upload to Radio**. Save an .img file of your factory settings immediately so you can always safely reset and recover from any digital mistakes.

**Bridge:** Cool. Now, can other humans in my house use this?

# Part 3:

## The Family Comms Plan (Simple, Repeatable, Stress-Proof)

Chapter 16

# The Family Comms Plan in One Page

In high-stress situations, human beings forget almost everything except the absolute simplest plan. Adrenaline is an incredible survival mechanism, but it completely destroys fine motor skills and complex memory recall. If cellular networks suddenly go down during a severe winter storm, a regional blackout, or a sudden neighborhood evacuation, your spouse or teenager will absolutely not remember how to navigate the Baofeng's VFO menu or manually program a CTCSS tone.

If they have to think about radio theory during an emergency, the plan has already failed. They need a communication strategy so painfully simple, so thoroughly rehearsed, that it fits entirely on a single index card.

**What you'll accomplish:**

- Create a one-page family communication plan that anyone in your household can follow under extreme stress.
- Assign clear, rigid roles so everyone knows exactly who is in charge of the radio network and who is checking in.
- Define strict check-in rules to conserve precious battery life and minimize airwave panic.

**What you need:**

- A notebook and a dark pen or marker.
- Your programmed Home Profile from Chapter 7.
- Your family members (or the core group of friends/neighbors you are planning with).

## The Three Rules of the Family Plan

Before we write a single frequency down, your family needs to sit at the kitchen table and agree on three non-negotiable rules for using the radios.

1. **Keep it Simple:** The radio stays permanently locked in Memory Mode. Nobody is allowed to press the menu button, change the squelch, or type random numbers on the keypad. The instruction should be exactly three steps: Turn the top knob on, use the arrows to pick the channel number, and press the side button to talk. That is it.
2. **Keep it Legal:** Emphasize heavily that these are not walkie-talkie toys for casual chatting. They are serious tools for coordinating when normal grid methods fail. These airwaves are public, and during a crisis, they are shared with official emergency personnel. Everyone must follow the licensing rules we discussed in the front matter and behave professionally.
3. **Keep it Practiced:** A plan is just a worthless piece of paper until you drill it into reality. Under stress, you don't rise to the occasion; you default to your level of training. We will cover engaging, 10-minute weekly drills in Chapter 19

to build this muscle memory.

## Assigning Roles

If an emergency strikes and everyone tries to grab a radio and talk over each other at the exact same time, you will have chaos. Assigning distinct roles creates a calm, military-style hierarchy.

- **Primary Operator (You):** You are the technical expert and the "quartermaster." You manage charging the batteries, programming the channels, and monitoring the wide-area repeaters for news and weather updates. You act as the central dispatcher.
- **Secondary Operator:** Usually a spouse, partner, or older teenager. This role is absolutely critical. If the Primary Operator is stuck at work across town or injured, the Secondary instantly takes complete control of the base station radio at home. They need to know exactly how to run the net.
- **Check-In Leads:** Anyone else holding a radio out in the field (kids at school, a partner commuting). Their only job is to turn the radio on, report their status to the Primary or Secondary operator at specific, pre-arranged times, and then turn the radio back off.

## When to Listen vs. When to Transmit

A Baofeng battery is a finite resource. It will happily last for about 12 to 18 hours if you are just quietly listening to traffic. However, pushing your voice out across a city takes a massive amount of physical power. If you hold down the PTT button and talk non-stop, a fully charged battery will die in less than two hours.

*STOP*

**The Walkie-Talkie Trap:** Do not treat the radio like an open phone call. If the grid goes down, the psychological urge to constantly chatter for comfort is incredibly high. You must resist this. Establish a firm household rule: The radio is for listening 95% of the time. You only transmit to deliver a scheduled check-in, acknowledge a received message, or report an immediate, life-threatening emergency.

## The "Say Less" Principle

When you do finally press the button to transmit, follow the "Say Less" principle. Under stress, people naturally tend to ramble. Rambling clutters the shared public channel, wastes precious battery power, and makes it impossible for the person on the other end to write down important details.

Instead of a breathless, rambling story about a flat tire and broken traffic lights, teach your family to use the highly structured **WHO / WHERE / WHAT / NEXT** format. It forces the brain to organize information before the mouth opens.

- **Who:** "Dad, this is Sarah." *(Establish identity immediately).*
- **Where:** "I am safe at the high school gym." *(Establish location).*
- **What:** "Roads are blocked by trees, I am staying here for now." *(Establish the situation).*
- **Next:** "I will listen for you at the top of the next hour. Over." *(Establish the future action).*

## Deliverable: The One-Page Template

Grab your notebook. Write this template down in bold, clear letters. You will eventually copy this onto a small, sturdy piece of paper, completely laminate it with clear packing tape to make it waterproof, and put it in your family's emergency kit (or tape it directly to the back of the radio's battery).

## FAMILY COMMS PLAN

- **Primary Operator:** *Name*
- **Secondary Operator:** *Name*
- **Home Channel (Simplex):** Channel 3
- **Backup Channel (Simplex):** Channel 4
- **City Info Channel (Listen Only):** Channel 1
- **Check-In Times:** Top of every hour strictly (e.g., 2:00, 3:00, 4:00).
- **Emergency Protocol:** If you are in immediate danger, say "Emergency, Emergency, Emergency" and give your location immediately.

*YOU'RE WINNING*

You just eliminated 90% of the sheer panic associated with suddenly losing cell service. Your family doesn't have to guess what to do. They now know exactly what channel to be on, who is supposed to talk, the format to use, and exactly what time to expect a reassuring message from you.

## Quick Win Checklist

- [ ] I understand why the family plan must be kept incredibly simple to survive an adrenaline dump.
- [ ] I have clearly assigned Primary and Secondary operator roles in my household.
- [ ] My family understands the "Walkie-Talkie Trap" and knows the radio is for listening 95% of the time to save battery.
- [ ] I have taught my family the "WHO / WHERE / WHAT / NEXT" format for clear, punchy messages.
- [ ] I have filled out the One-Page Family Comms Plan template and prepared to laminate it.

## If you're stuck, go to...

- *Chapter 8: Troubleshooting* (If a family member accidentally unlocks the radio keypad, changes a setting, and suddenly can't hear you, run them through the 60-second diagnostic flow).

Chapter 17

# Simplex That Works (What You Can Expect and Why)

Hollywood movies and aggressive walkie-talkie marketing boxes have lied to us. In the movies, a hero with a cheap handheld radio talks to a rescue helicopter three states away, achieving crystal-clear audio from inside a concrete basement. On the box of bubble-pack radios at the sporting goods store, you'll see bold claims like "36-MILE RANGE!"

In reality, pure radio-to-radio communication (referred to as "Simplex" or "Direct") is strictly bound by the unyielding laws of physics. If you hand your spouse a Baofeng and expect them to hear you clearly from ten miles away across a dense, hilly city without using a repeater tower, you are going to be severely disappointed. Understanding your actual limitations is the secret to building a plan that won't fail you. Let's set highly realistic expectations so your family doesn't panic the first time they hear a wall of static.

**What you'll accomplish:**

- Set realistic, field-tested range expectations for handheld radios.
- Learn how to instantly improve your results using location, environment, and body positioning.
- Choose simple, reliable family channels that actually work when the grid goes down.

**What you need:**

- Two programmed Baofeng radios.
- A partner to help you test.

## Range Reality: The Line-of-Sight Rule

Your Baofeng transmits in the VHF (Very High Frequency) and UHF (Ultra High Frequency) bands. These radio waves behave almost exactly like a beam of light from a powerful flashlight.

If you shine a flashlight at a dense forest, the light scatters against the tree trunks and quickly stops. If you shine it at a concrete building, the light bounces off the wall. If you stand on top of a mountain and shine it down into a valley, the light travels unhindered for miles. Your radio waves act the exact same way.

- **Dense Urban City:** Expect 0.5 to 1 mile of reliable range. Steel-framed buildings, thick concrete walls, and massive electrical interference (the invisible "RF smog" created by thousands of Wi-Fi routers and power lines) will eat your 5-watt signal alive.

- **Suburban Neighborhood:** Expect 1 to 3 miles of range. Wood-framed houses are easier to punch through, but rolling hills and dense trees scatter the signal. Interestingly, you will often get slightly better range in the winter when the leaves have fallen off the trees, compared to the thick, wet foliage of summer.
- **Flat Open Country / Over Water:** Expect 3 to 5 miles of range. With nothing in the way to absorb the signal, your radio will reach out until the physical curvature of the earth literally gets in the way of the beam.
- **Mountaintop to Valley:** Expect 10 to 30+ miles of range. Height is absolutely everything in VHF/UHF radio. If your antenna can visually "see" the other antenna, the signal will make the trip.

## Where Simplex Shines

If the range is so seemingly limited compared to a cell phone, why do we use Simplex channels at all? Because Simplex is entirely **grid-independent**.

If a massive winter storm knocks out power to your entire county, the cellular towers will go down within hours. A few hours later, the backup diesel generators running the local ham radio repeaters might run out of fuel. When absolutely all external infrastructure fails, Simplex is the only thing left. It requires nothing but your battery and their battery. It is absolutely perfect for:

- Coordinating between two cars in a convoy traveling through rural areas with zero cell service.
- Checking in with a spouse who is walking home from a neighborhood grocery store during a blackout.
- Running perimeter security around a rural property, farm, or campsite.

## How to Instantly Improve Your Signal

If your spouse calls you on your designated Simplex channel and their voice is suddenly buried in scratchy, popping static, do not tell them to repeat themselves immediately. Tell them to change their physical environment first.

- **The "Step Outside" Rule:** Standard wood walls absorb about 30% of your radio signal. Brick, stucco with metal wire mesh inside (which acts like a giant cage), and modern energy-efficient windows with metallic coatings can absorb up to 80% of your signal. Simply walking out the front door, or stepping onto a wooden back porch, can turn a terrible, broken signal into a crystal-clear one instantly.
- **Get High:** Remember the flashlight analogy. Standing on a second-story balcony or at the top of a staircase is always vastly superior to transmitting from a ground-floor kitchen or a basement.
- **The "Two-Foot Shift":** UHF waves are incredibly short physically (about 70 centimeters long). Sometimes, radio waves bounce off a parked car or a metal fence and collide with each other, creating an invisible "dead spot" right where you are standing. Moving your body just two feet to the left or right can pull your antenna completely out of that dead zone and restore the audio.

## Mini-Experiment: The Neighborhood Walk

It's time to prove these concepts in the real world. By doing this now, you eliminate the fear of the unknown later.

**Step 1:** Give one radio to your partner. Have them sit at the kitchen table (the base station).

**Step 2:** Ensure both radios are locked onto Channel 3 (your Family Simplex channel from Chapter 7) and the volume is turned up to a comfortable level.

**Step 3:** Take your radio and start walking down the street, away from your house.

**Step 4:** At 1/4 mile, press the PTT button. Pause for one second. Say: "Base, this is *Name*. I am at the corner of 4th and Main. How is my signal? Over." Have them reply. Both of your voices should be perfectly clear.

**Step 5:** Walk to 1/2 mile. Intentionally stand behind a large brick building, or walk down into a geographic dip in the road. Make the exact same call. Notice how the "hiss" of static starts creeping into the background of their voice.

**Step 6:** Now, step out from behind the building, or walk 50 feet up to the highest point on the street, and call again. Notice how the signal instantly clears up just by altering your line-of-sight.

*FIELD NOTE*

**Build a Mental Map:** Writing down the results of this walk on a piece of scratch paper gives you a permanent "mental map" of your neighborhood's radio terrain. If an emergency happens, you won't have to guess. You will already know exactly which street corners are total dead zones and which nearby hills are the perfect spots to stand when you need to transmit a critical message back to the house.

## Quick Win Checklist

- [ ] I understand that VHF/UHF radios are strictly "line-of-sight" devices, not magic wands.
- [ ] I have highly realistic expectations for my neighborhood range (usually 1 to 3 miles).
- [ ] I understand that Simplex is a vital lifeline because it requires absolutely zero grid infrastructure to work.
- [ ] I know to step outside, get higher, or shift my position by two feet if an incoming signal is scratchy.
- [ ] I have conducted a "Neighborhood Walk" with my partner to map out the dead zones and high points around my home.

## If you're stuck, go to...

- *Chapter 9: I Can Hear Them, But They Can't Hear Me* (if you do the neighborhood walk and you hear your partner perfectly, but they hear absolutely nothing but silence when you reply).

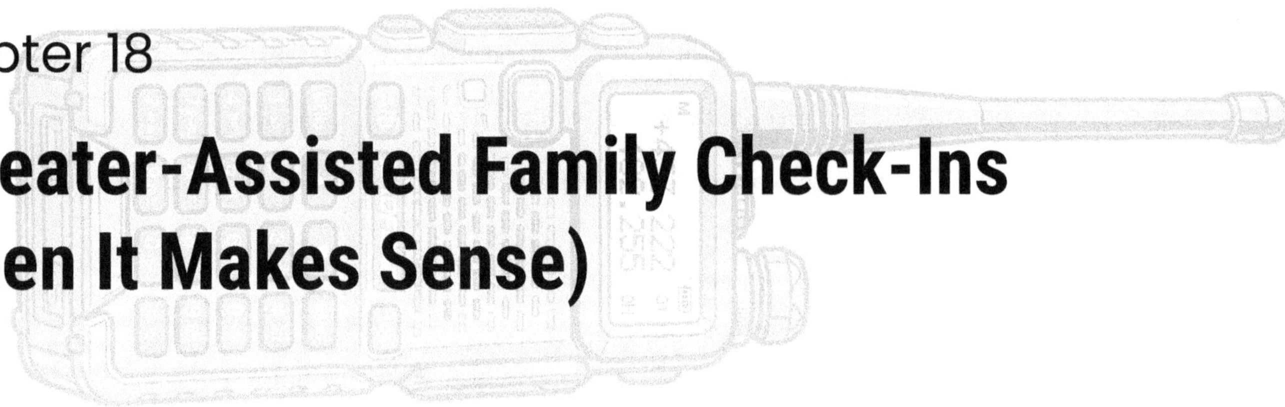

Chapter 18

# Repeater-Assisted Family Check-Ins (When It Makes Sense)

SIMPLEX IS FANTASTIC FOR your immediate neighborhood, but what if you work ten miles away across the city in a concrete office park? What if your teenager's high school is on the other side of a massive, heavily wooded hill? If the cellular network suddenly crashes while your family is scattered during a severe weather event, pure radio-to-radio communication simply isn't going to punch through that much terrain and steel.

You need a bigger antenna, and you need a massive altitude advantage. This is where we bring the public repeater concepts you learned back in Part 1 directly into your family's practical emergency plan.

**What you'll accomplish:**

- Determine exactly how and when your family actually needs a repeater for emergency check-ins.
- Designate one official "Family Repeater" to act as your long-range rally point.
- Build a strict, disciplined schedule that works around the busy nature of public radio towers.

**What you need:**

- Your Baofeng radio.
- Your One-Page Comms Plan from Chapter 16.

## When Repeaters Help (And When They Don't)

As we learned earlier, repeaters are the ultimate cheat code for handheld radios. They act like towering lighthouses, taking your weak, 2-mile signal from the ground and blasting it out over a massive 30-mile radius. However, when relying on them for survival communication, you must understand their one fatal vulnerability: **They belong to the grid.** These massive antennas sit on top of commercial buildings, water towers, or mountain peaks, and they run on standard commercial power. If a hurricane, earthquake, or severe winter freeze knocks out regional power, the repeater tower will automatically switch to a backup system. It might have a battery bank. It might have a diesel generator. It might have solar panels.

A good diesel generator might keep the tower alive for three days. A solar setup might struggle on day two of a dark blizzard. But eventually, if the grid stays down long enough, the repeater will inevitably die.

- **The Golden Rule of Thumb:** Use a repeater exclusively to coordinate getting your family home *during the first few hours* of a crisis. This is the "golden window" when anxiety is highest and people are actively trying to navigate blocked roads. Once everyone is safely back at the house, or within a 2-mile radius of each other, immediately switch off the repeater and use your private Simplex channels to conserve the tower's resources.

## Choosing the "Family Repeater"

In Chapter 7, we saved two powerful repeaters into your Home Profile (Channel 1 and Channel 2). You now need to pick one of those to be your officially designated family rally point.

1. **Pick the tested, reliable one:** Do not just pick the repeater that is physically closest to your house. Pick the one that gave both you *and* your spouse the clearest, loudest signal during your tests in Part 1. A repeater on a mountain 15 miles away might actually perform better for both of your commute routes than a smaller repeater just 3 miles down the road.

2. **Make it official:** Write this specific channel clearly on your One-Page Comms Plan under "City Info Channel." Everyone in the family needs to know that this is the designated meeting spot in the sky.

3. **The Backup:** Technology fails. Antennas snap in high winds. If your primary repeater gets knocked offline by a storm and all you hear is dead static when you transmit, your family must automatically know to click up to Channel 2 (The Backup Repeater) and try to connect there.

## The Check-In Schedule (Timing is Everything)

Here is the hard, unavoidable truth about using a repeater: **You do not own it.** It is a shared, public resource built and maintained by local amateur radio clubs using their own time and money.

During a major, wide-area emergency (like a spreading wildfire, a tornado touchdown, or a massive flood), official emergency volunteer nets (often called ARES or RACES) will immediately activate. They will take complete control of the primary repeaters to coordinate critical information with local fire departments, police, and hospitals.

**[STOP] Stay Off the Button:** If official emergency traffic is happening, you must stay off the button. Do not interrupt an active rescue operation just to tell your spouse you are stuck in traffic.

To work around this public shared space, your family check-in schedule needs to be incredibly precise and highly disciplined.

- **Top of the Hour:** In emergencies, staring at a silent radio waiting for a call will drive you crazy. Families should agree to only turn their radios on and attempt to contact each other exactly at the top of the hour (e.g., 1:00, 2:00, 3:00).

- **The 3-Minute Window:** This eliminates the anxiety of waiting. Both you and your spouse turn your radios on to Channel 1 at exactly 1:58 PM. You just listen. If the channel is clear of emergency traffic, you make your call at exactly 2:00 PM. By 2:03 PM, the check-in window is officially over, and you turn the radios completely off to save your batteries for the next hour.

*GOOD ENOUGH*

What if the repeater is totally busy with official emergency traffic at 2:00 PM? Don't panic and don't interrupt. You simply listen quietly to the traffic, which often provides incredible, life-saving clues about the disaster, like which roads are flooded or closed, and then try to make your family contact at the bottom of the hour (2:30 PM) instead.

## The Quick-Strike Script

When you use a public repeater during a crisis, hundreds of other families might be waiting to do the exact same thing. A five-minute conversation about your flat tire or how scared you are is incredibly selfish and potentially dangerous to others waiting to speak. You must use the absolute shortest script possible.

*(Wait for the channel to be totally clear of talking...)* "This is *Your Call Sign* attempting to reach *Spouse's Call Sign*. Are you copying?" *(Spouse replies...)* "This is *Spouse's Call Sign*. I am safe at the office. Leaving in one hour. Over." "Copy that. See you at home. *Your Call Sign* clear."

Ten seconds. That is all it takes. You got the vital location information, you stayed legally identified with your call signs, and you immediately got out of the way so other people could use the tower.

## Quick Win Checklist

- [ ] I understand that repeaters are vulnerable to long-term power outages and should be used primarily to get people home quickly during the initial hours of a crisis.
- [ ] I have officially designated Channel 1 (the most thoroughly tested tower) as my Family Repeater rally point.
- [ ] My family understands they must automatically switch to Channel 2 if Channel 1 is offline or unreachable.
- [ ] I understand that repeaters are public, and I must never interrupt official emergency or rescue nets.
- [ ] I have set a strict "Top of the Hour / 3-Minute Window" check-in schedule to conserve battery life and limit airtime.

## If you're stuck, go to...

- *Chapter 11: I Can Hit One Repeater But Not Another* (If you successfully designate a Family Repeater, but your spouse can't seem to hit it with their Baofeng from inside their concrete office building, run the troubleshooting steps).

# Narrative Interlude: Callsign Day

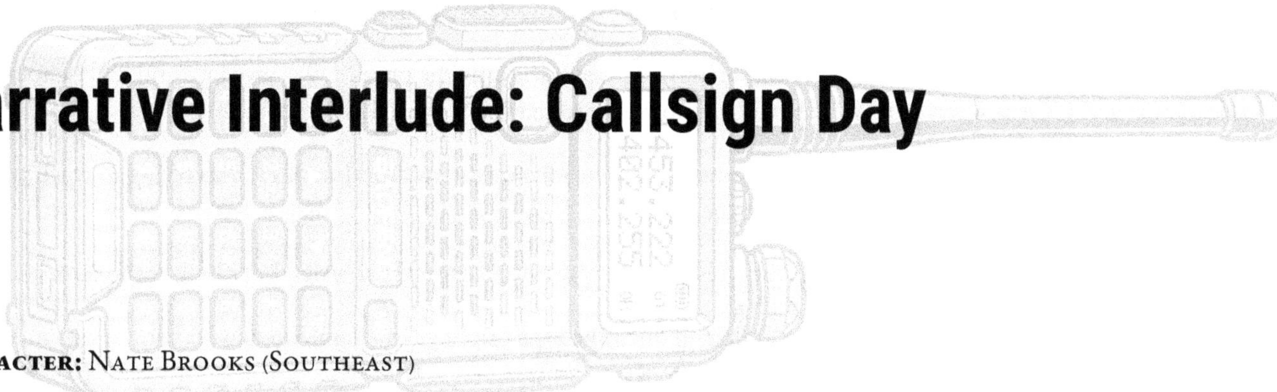


**Character: Nate Brooks (Southeast)**

**Skills reinforced:**

- One-page family plan
- Simplex expectations
- Repeater-assisted check-ins
- Short message format (WHO/WHERE/WHAT/NEXT)

**Where to learn the steps:** Chapters 16, 17, & 18

The email arrived at 7:14 AM on a Thursday. Nate Brooks was sitting at his kitchen island, sipping black coffee. He opened the FCC database link on his phone, typed in his last name, and there it was.

**KN4XYZ.**

It was a quiet moment, but Nate felt a distinct surge of serious pride. The letters and numbers weren't just a license; they were a certification of competence. He had done the reading, passed the test, and secured his legal right to transmit. He was fully operational.

Kara walked into the kitchen, tossing her keys onto the counter. "You've got that look," she said, eyeing him as she poured her own coffee. "The 'mission accomplished' look."

"Callsign day," Nate said, turning the phone screen toward her. "It's official. The grid-down comms plan is fully legal and ready to execute."

"Great," Kara said, taking a sip and looking him dead in the eye. "Now make this usable. Because right now, those radios are just intimidating plastic bricks sitting on your workbench."

Nate nodded. He knew she was right. A system is only as good as its weakest operator. That evening, after dinner, he called a mandatory family meeting in the living room. Kara sat on the sofa. Their sixteen-year-old daughter, Leah, slumped into the armchair, her eyes glued to a TikTok video playing silently on her phone.

Nate stood in front of the coffee table, where he had laid out three programmed Baofeng UV-5Rs.

"Alright," Nate began, slipping effortlessly into his professional briefing voice. "As we approach peak hurricane season, we need to establish our primary and secondary communication protocols. In the event of catastrophic cellular failure, these radios operate on VHF and UHF frequencies. Because UHF is strictly line-of-sight, we have to account for terrain masking and structural absorption. Now, if we need to reach across the county, we will utilize a duplex offset through the primary repeater, which requires a digital tone, "

"Dad," Leah interrupted, not looking up from her screen. "I have literally no idea what you are saying."

Nate stopped. He looked at Kara. She was tapping her foot, offering a thin, patient smile.

"You're lecturing, Nate," Kara said gently. "You spent three weeks learning the physics. We have jobs and high school. We are not passing an FCC exam tonight. We just need to know how to call you if the lights go out."

Nate looked at the radios, then at his family. He had fallen into the classic technical trap. He was trying to teach them *how* the watch was built, instead of just telling them the time.

"You're right," Nate said. He took a deep breath, resetting his posture. "Forget the physics. Let's look at the card."

He reached over to the primary base station radio and unclipped the bright yellow, laminated index card he had made in the garage. He handed a duplicate card to Leah, and one to Kara.

"This is the One-Page Comms Plan," Nate said, his voice dropping the rigid, tactical cadence and becoming calm and conversational. "Everything you need to know is on this yellow card."

Leah finally put her phone down and looked at the card.

"Rule one: Roles," Nate said. "I am the Primary Operator. If I am not home, Kara, you are the Secondary. You run the net from this house. Leah, your only job is to turn the radio on, check in, and turn it off."

"Okay," Leah said, sitting up slightly. "I can do that. Which button is it?"

"Top knob," Nate said. "Turn it until it clicks. That's it. Now, look at the channels on the card. Channel 3 is our Simplex channel. That means direct, radio-to-radio. It doesn't rely on cell towers or power grids. But there is a catch: it only works within a mile or two. If you are at the neighborhood entrance, it works perfectly. If you are downtown, it won't reach."

"So what do we do if we're downtown?" Kara asked, tracing the card with her finger.

"You click the arrow up to Channel 1," Nate explained. "That's the City Repeater. It's a massive tower that bounces our signals 30 miles. But, because it's public, it will be incredibly busy during an emergency. We cannot clog it up. So, we follow the Check-In Rule."

He pointed to the bottom of the yellow card.

"Top of the hour," Nate said firmly. "We do not sit around staring at silent radios. If the cell phones die, you turn your radio on at exactly 2:00 PM, 3:00 PM, or 4:00 PM. We have a three-minute window. We make contact, we exchange status, and we turn the radios off at 2:03 PM to save battery."

"What do I say?" Leah asked, looking at the intimidating keypad on the front of the radio. "Do I have to use police codes?"

"No codes," Nate said. "Plain English. Look at the back of the card. It's four words: Who, Where, What, Next."

He picked up a radio. "It sounds exactly like this: *Dad, this is Leah. I am at the high school. We are sheltering in place. I will check back in one hour. Over.*"

He set the radio down. "That is the entire system. Who you are, where you are, what your situation is, and what your next move is. Ten seconds, and you're off the air."

Kara nodded slowly. "Roles. Channel 3 for local. Channel 1 for long distance. Top of the hour. Who, where, what, next." She looked at Nate. "That's a plan I can actually remember."

"Let's prove it," Nate said. He looked at his watch. It was 7:45 PM. "Leah, take your radio. Walk down to the neighborhood clubhouse. At exactly 8:00 PM, I want you to initiate a Simplex check-in on Channel 3 using the four-step format."

Leah grabbed the radio, slipping the yellow card into her pocket. "Top of the hour. Got it."

For the next fifteen minutes, Nate and Kara sat in the kitchen. Nate forced himself not to touch his base station radio. At 7:59 PM, he turned the top knob on and verified the screen was on Channel 3. He set it on the island counter.

At exactly 8:00 PM, the squelch gate snapped open, accompanied by a faint hiss of static, proving they were operating on direct, line-of-sight simplex.

*"Dad, this is Leah,"* her voice came through, completely clear. *"I am at the clubhouse. The pool is closed. Heading back to the house now. Over."*

Nate smiled. He picked up the radio, pressed the PTT button, and paused for one second.

*"Copy that, Leah. Loud and clear. See you in five minutes. KN4XYZ clear."*

He released the button and looked at Kara.

"She did it perfectly," Kara said, genuinely impressed. "No panic. No confusion."

"Because she didn't have to guess," Nate said, tapping the laminated yellow card on the counter. He had learned the most valuable lesson of emergency preparedness: the best gear in the world is useless if the procedure is too complex for a tired, stressed-out human being to remember.

He had his callsign. He had his hardware. But more importantly, he finally had a system his family could trust.

## Takeaway

A plan that can't be explained in one page won't be followed. Do not burden your family with radio theory, physics, or complex menu operations. Give them a highly structured, laminated cheat sheet. Establish clear roles, realistic Simplex range expectations, strict check-in times to conserve battery, and enforce the short WHO/WHERE/WHAT/NEXT message format to eliminate rambling.

**Bridge:** Now we build the habit: 10 minutes a week.

Chapter 19

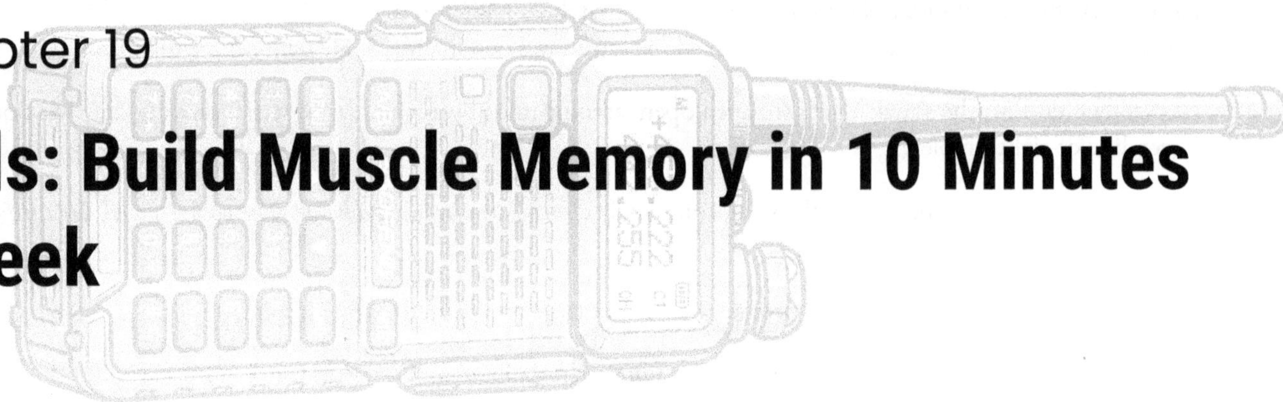

# Drills: Build Muscle Memory in 10 Minutes a Week

A BEAUTIFUL, PERFECTLY THOUGHT-OUT plan written on a laminated index card is completely useless if nobody in your household actually knows how to execute it under pressure.

If the very first time your spouse or teenager tries to turn on the Baofeng and change a channel is during a terrifying neighborhood evacuation in the pitch black, it will be a complete disaster. Adrenaline is a powerful chemical that causes tunnel vision and destroys fine motor skills. Under that kind of intense stress, their hands will shake, their brain will freeze, they will press the wrong buttons, the radio will beep angrily at them, and they will give up in frustration, assuming the radio is broken.

Muscle memory is the only thing that reliably defeats panic. You don't rise to the occasion during a crisis; you default to your level of training. We need to turn operating this radio into a mindless, boring, automatic habit.

**What you'll accomplish:**

- Create highly repeatable, 10-minute drills for your family that don't feel like a chore.
- Turn basic radio operation into a stress-free, automatic habit that survives an adrenaline dump.
- Use gamification (like the "Lost Phone" drill) to keep kids and teenagers engaged in the process.

**What you need:**

- Your family members.
- Your programmed radios (fully charged).

## Drill 1: The Blindfold Test (Basic Hardware)

In a real emergency, like a severe winter storm or a tornado, the lights are usually the first thing to go out. Before anyone worries about talking on the airwaves, they need to know how to manipulate the physical hardware without staring at the screen. You should be able to operate the basic functions of this radio in total darkness.

**The Drill:**

1. Hand the radio to your family member completely turned off and with the volume knob rolled all the way down.
2. Ask them to close their eyes (or sit together in a completely dark room).

3. Say: *"Turn the radio on, set the volume to a low, comfortable level, and prove the speaker actually works."*

4. **Pass Condition:** Without opening their eyes, they successfully click the top knob on. They physically slide their fingers down the left side of the radio to find the MONI button (distinguishing it from the larger PTT button by feel), press it to hear the rushing static, adjust the volume knob to a comfortable level, and release the button.

## Drill 2: The Channel Switch

Under stress, people tend to grip the radio tightly, accidentally crushing the keypad with the palm of their hand. This drill ensures they can reliably jump between your City Repeater and your Family Simplex channel without accidentally pressing the wrong keys and getting lost in the menus.

**The Drill:**

1. Put the radio in Memory Mode on Channel 1 (with the keypad officially locked).

2. Say: *"The city repeater just went down and all we hear is static. Switch to the Backup Family Channel."*

3. **Pass Condition:** They calmly hold the # key for two seconds until the tiny padlock icon disappears, use the Up Arrow to click up exactly three times to Channel 4, and press the # key again to re-lock the pad.

*STOP*

**Correct gently.** If they accidentally press the orange VFO/MR button and suddenly get lost in frequency mode, do not yell or get frustrated. Panic kills the desire to learn. Calmly remind them: "Take a breath. If you see a screen full of raw numbers without a tiny channel number on the right, you just accidentally bumped the door open. Press the orange button once to get back home."

## Drill 3: The "Clean Call"

Now we practice the "Say Less" principle from Chapter 16. It is incredibly common for beginners to start talking *before* they press the button, which cuts off the first half of their sentence.

**The Drill:**

1. Put one family member in the living room with a radio on Channel 3. Go out to the backyard with your radio so you can't see each other.

2. Text them on their phone: *"Initiate a check-in."*

3. **Pass Condition:** They must execute the perfect "Push, Pause, Speak" rhythm. They press the PTT button, pause for a deliberate half-second, state who they are, state where they are, state their status, and cleanly say "Over," then immediately release the button. *(Good Example: "Dad, this is Leo. I am in the living room. Everything is secure here. Over.") (Bad Example: "...ead, it's Leo, I'm just sitting here on the couch, what do you want me to do next?")*

## Drill 4: The "Lost Phone" Scenario (Advanced)

Teenagers might roll their eyes at sitting in the living room practicing radio calls. Once a month, run a practical, gamified scenario that simulates a minor crisis in the real world. This proves to them that the tool actually works when the cell network doesn't.

**The Drill:**

1. Tell your family: *"For the next 30 minutes, all cell phones are dead."* Have everyone physically put their smartphones in a drawer.
2. Send a family member to the neighborhood grocery store, a friend's house, or a nearby park (somewhere you have already proven is within your Simplex range) with a Baofeng radio.
3. Have them coordinate their pickup, or report when they are walking home, using *only* the radio on the strict Top of the Hour check-in schedule.
4. **Pass Condition:** They successfully contact you at the designated time, deliver the required information cleanly without rambling, and return home. The feeling of success when they hear your voice without using a cell tower is a massive confidence booster.

## Tracking the Habit

Consistency beats intensity every single time. Don't try to run all four drills on the same day for an hour straight. People will get incredibly bored, annoyed, and view the radios as a chore.

Run just one drill for ten minutes on a Saturday morning while making coffee. Keep a simple physical checklist on the fridge. Once everyone in the household can pass Drill 1 and Drill 2 flawlessly without hesitating or looking at the buttons, your family is officially better prepared than 99% of your neighborhood for a local emergency.

**[YOU'RE WINNING]** You have successfully transitioned your family from passive bystanders to capable radio operators. The panic is gone.

## Quick Win Checklist

- [ ] I understand that muscle memory is the only reliable tool during a stressful emergency or adrenaline dump.
- [ ] I have successfully run the "Blindfold Test" with my family members to ensure they can find the volume and monitor buttons by feel alone.
- [ ] I have verified my family can confidently unlock the keypad, change channels precisely, and re-lock it.
- [ ] I have practiced "Clean Calls" to enforce the Push-Pause-Speak rhythm and the WHO/WHERE/WHAT protocol.
- [ ] I have planned one simple, 10-minute radio drill for this upcoming weekend to start building the habit.

## If you're stuck, go to...

- *Chapter 12: My Channels Won't Save* (If a family member accidentally deletes a channel while fumbling with the keypad during a drill, go back and rebuild it using the Double Save method).

Chapter 20

# Go-Bag Setup (Power, Spares, and What Actually Matters)

You wake up at 2:00 AM to the terrifying wail of tornado sirens. The power grid in your entire neighborhood goes completely black. Your heart rate spikes. You calmly reach into the hallway closet, grab your trusty Baofeng radio, twist the volume knob on, fully expecting to hear the comforting voice of the local weather dispatcher... and the screen stays black.

No beep. No static. Nothing. The lithium battery slowly died three months ago while sitting neglected in the bottom of a drawer.

A radio without a rock-solid power and storage strategy isn't a lifeline; it's just a tactical-looking plastic brick. In an emergency, you will not have the time, the electricity, or the presence of mind to hunt for a charging cable. Preparedness means ensuring your equipment is in a permanent, organized "ready state," so it functions perfectly the exact second you reach for it in the dark. Let's cut through the internet hype and build a foolproof, budget-friendly Go-Bag setup that actually works when the lights go out.

**What you'll accomplish:**

- Develop an automated power strategy so your radio is never dead when you need it most.
- Understand which antenna upgrades actually matter (and which "tactical" traps are a waste of money).
- Build a consolidated, grab-and-go kit for your emergency bag that protects your gear.
- Learn the secret to safely storing a backup radio in your vehicle without ruining it.

**What you need:**

- Your Baofeng radio.
- The original desktop charging cradle.
- A dedicated small padded pouch or bag.

## The Power Strategy (Rule of Two)

Lithium-ion batteries are fantastic, but they have a fatal flaw: they naturally lose their charge over time, even when the radio is completely turned off. Furthermore, the Baofeng has a tiny amount of "parasitic drain", meaning the internal circuitry sips a microscopic amount of power even when the knob is clicked off. If you fully charge your radio in January and put it in a closet, it might be completely dead by July.

- **The Rotation (Avoid the Silent Killer):** Never, ever store the radio sitting inside the plugged-in charging cradle. The Baofeng charging docks are famously "dumb." They constantly trickle a tiny bit of voltage into the battery, which degrades the chemical cells over time and can cause the battery pack to physically swell or fail. Instead, charge the battery to 100%, take it entirely off the charger, and put it in your bag. Set a recurring calendar reminder on your phone for the 1st of every month to plug it in and "top off" the charge for an hour.
- **The Rule of Two:** "Two is one, and one is none." You must own at least one spare, extended battery (like the incredibly popular 3800mAh extended battery pack). The extended battery makes the radio slightly taller and heavier, but it effectively doubles your listening time from 14 hours to almost 30 hours. Keep the extended battery physically attached to the radio, and keep the smaller stock battery fully charged sitting right next to it in the bag. Use a silver marker to write A and B or the date of the last charge on them.
- **The AA Battery Shell (The Ultimate Grid-Down Backup):** You can buy a special, empty plastic battery shell designed for the UV-5R that holds 6 standard AA batteries. If the power grid is down for a massive two-week blackout and you absolutely cannot plug your charger into a wall, popping in standard AA batteries from your kitchen drawer is an absolute lifesaver.

*STOP*

**Not all AAs are created equal.** Do not put standard alkaline batteries (like cheap Duracells or Energizers) into your backup shell and leave them in a hot bag for three years. They will leak corrosive battery acid and permanently destroy the plastic shell. You must use Lithium AA batteries (like Energizer Ultimate Lithium). They cost twice as much, but they have a 20-year shelf life, they perform beautifully in freezing weather, and they absolutely never leak.

## Antennas: Keep It Simple and Avoid the "Tactical" Trap

New radio operators love to jump on Amazon and immediately buy massive, 15-inch "tactical" whip antennas that fold in half with a velcro strap. They look incredibly cool, but they introduce a massive vulnerability.

While these massive antennas *do* slightly improve your transmit range, they make the radio incredibly top-heavy and clumsy to hold. More importantly, they create leverage. If you drop the radio on the kitchen floor, or if the long antenna snags on a car door as you are getting out, that 15-inch lever will instantly snap the fragile brass SMA connector completely off the internal circuit board. Your radio is now effectively useless and requires complex micro-soldering to repair.

- **The Go-Bag Standard:** The 6-inch stock "rubber duck" antenna that came in the box is perfectly fine for 90% of neighborhood emergencies. It is short, flexible, and extremely difficult to break. Keep it on the radio.
- **The Smart Upgrade:** If you desperately want a range upgrade, skip the heavy tactical folding whips. Look for a "Signal Stick" (made of a special memory-metal that can be tied in a knot without breaking and snaps right back into perfect shape) or a genuine Nagoya NA-771. However, if you attach an upgraded antenna, you must keep the original stock antenna tucked away in your bag as a permanent backup. Antennas break easily in chaotic environments, and a radio without an antenna is just a plastic brick.

## The Grab-and-Go Kit

Do not leave your radio components scattered around the house. The charger is plugged into the wall in the home office, the radio is sitting on the kitchen counter, and the spare battery is buried in a junk drawer. When the lights violently go out and panic sets in, you will never find all three pieces in the dark.

Buy a small, inexpensive padded electronics pouch, a tactical admin pouch, or a small Pelican-style hard case. Inside this single, dedicated container, pack:

1. **The Radio:** With the extended battery and stock antenna physically attached, ready to turn on.
2. **The Spare Battery:** Fully charged.
3. **The AA Battery Shell:** Pre-loaded with 6 Lithium AA batteries.
4. **Your Comms Plan:** The printed, waterproofed, laminated One-Page template from Chapter 16.
5. **A Discreet Earpiece:** In an emergency shelter or a crowded room, you might not want a loud dispatcher's voice broadcasting news to everyone around you. A cheap earpiece provides critical privacy and prevents panic from spreading. It also prevents drawing unnecessary attention to your supplies if you are trying to remain low-profile.
6. **A Headlamp & Pencil:** A small, cheap LED headlamp (so you can read the radio screen and use both hands to operate it in the dark) and a small carpenter's pencil (which will never dry out or freeze like an ink pen) to log important information.

## The "One Extra Radio" Concept

Baofengs are incredibly cheap, often less than the cost of a large pizza. Once you have built your perfect "Home Profile," run your family drills, and packed your Go-Bag, the absolute smartest thing you can do is buy one more radio.

Program this second radio exactly like yours. Turn it off, lock the keypad, and put it in the glovebox of your car or the bottom of your spouse's daily work bag.

If an emergency happens while you are physically separated at work or running errands, you don't have to stress about whether or not they remembered to bring their radio today. You already know they have one, you know it's programmed with the exact same frequencies, and you know exactly which channel to find them on.

*FIELD NOTE*

**Surviving the Hot Car:** If you decide to store a backup radio in your vehicle, do not leave the lithium-ion battery attached to it. The interior of a car parked in the summer sun can easily reach 140°F (60°C). Extreme heat will completely destroy a lithium battery, causing it to swell or even catch fire. For a dedicated vehicle radio, remove the lithium battery entirely. Power it exclusively using the AA Battery Shell loaded with Energizer Ultimate Lithium AA batteries, which are chemically designed to safely withstand extreme temperature swings without degrading or exploding.

*YOU'RE WINNING*

Take a moment to look at your organized, fully charged, programmed kit. You have successfully graduated. You started this book with a confusing, beeping, stressful piece of plastic, and you have methodically transformed it into a trusted, tested, family-integrated emergency lifeline. You are officially ready for whatever the world throws at you.

## Quick Win Checklist

- [ ] I understand the danger of parasitic drain and have set a monthly calendar reminder to top-off my battery charge.
- [ ] I have permanently removed my radio from the desktop charging cradle so I don't degrade the battery chemistry over time.
- [ ] I have avoided the "tactical antenna" trap to protect my radio's fragile antenna connector.
- [ ] I have consolidated my radio, spare battery, earpiece, and Comms Plan into a single, easy-to-grab padded pouch.
- [ ] I understand how to safely store a backup radio in a hot vehicle using Lithium AA batteries.
- [ ] I have completely finished Part 3! My family is fully prepared, practiced, and equipped to communicate in an emergency.

## If you're stuck, go to...

- *Part 4: Quick Cards and Templates* (Now it's time to print out the final, concentrated cheat sheets so you have all of this critical information securely on paper for your Go-Bag).

# Narrative Interlude: Ten Minutes a Week

**Character:** Mark Jensen (Midwest Suburbs)

**Skills reinforced:**

- Weekly drills
- Go-bag readiness
- Calm check-ins
- Labeling/ready state

**Where to learn the steps:** Chapters 19 & 20

The local meteorologist on the morning news was pointing to a massive, angry red blob sweeping down from Canada. It was a classic Minnesota "Alberta Clipper," promising white-out conditions, plummeting temperatures, and a very high probability of downed power lines by nightfall.

Mark Jensen stood at the kitchen island, absolutely vibrating with preparedness.

Spread across the granite countertop was a tactical, olive-drab backpack with a dozen molle webbing straps. Next to it sat three Baofeng UV-5R radios, a tangle of extended antennas, battery charging cradles, a throat microphone he'd ordered late one night on Amazon, and a notebook filled with tiny, scribbled frequency tables.

Jen walked into the kitchen, holding her empty coffee mug. She stopped dead in her tracks, staring at the military-grade garage sale covering her counter. Owen shuffled in behind her, wearing pajama pants and earbuds.

Jen looked at the red blob on the TV, and then looked at Mark. She gave him *the look*. It was a look of deep, loving caution that explicitly meant: *Do not make this complicated.*

"The storm is tracking south," Mark announced, adopting his best Mission Control voice. "We have a highly probable grid-down scenario. I have spent the morning establishing a multi-tiered communications web. I've programmed fourteen local repeaters, a cross-band simplex relay, and I've configured dual-watch so we can monitor county fire dispatch while simultaneously tracking the weather node."

Owen slowly pulled one earbud out. "Dad, are we going to war with the snow?"

"No, Owen, we are ensuring continuity of operations," Mark said, picking up the throat microphone. "Now, for today's drill, I want us to practice switching between the primary and tertiary repeater offsets while wearing these tactical headsets. It will take about an hour, "

"Mark," Jen interrupted gently, sliding her coffee mug onto a small sliver of empty counter space. "You're doing the NASA thing again."

"I am establishing a robust protocol," Mark protested mildly.

"You have a throat mic," she pointed out. "I am not wearing a throat mic to tell you the power went out. You spent two weeks making that beautiful, simple yellow card. Why are we suddenly trying to run a CIA safehouse?"

Mark looked down at the sprawling mess on the counter. He looked at the giant, floppy 15-inch tactical antennas he'd bought. He looked at the charging cradles that were slowly degrading his lithium batteries. He realized she was entirely right. In his excitement to be "ready," he had completely abandoned the core principle that got him here: *Keep it simple.*

He had fallen back into the trap of thinking more gear meant more safety.

"Okay," Mark sighed, a smile breaking through. "You're right. I'm over-engineering. Let's dial it back."

Mark took the throat mic and the massive antennas and shoved them back into a cardboard box. He took the radios completely off their charging cradles, remembering the 'parasitic drain' warning from the guide, and placed them in a neat row.

He pulled a silver metallic Sharpie out of the kitchen drawer. On the back of the first radio's battery, he wrote **BASE**. On the second, he wrote **JEN**. On the third, he wrote **OWEN**.

"Labeling. Good," Jen nodded approvingly. "What's the actual plan?"

"Ten minutes. Two drills. That's it," Mark said, pointing to the laminated yellow Quick Card taped to the Base radio. "Owen, you're up first. Drill 1. The Blindfold Test."

Owen rolled his eyes, but he put his phone down. "I don't have a blindfold."

"Just close your eyes," Mark said, handing Owen his labeled radio. "The power is out. It's pitch black in the house. The cell networks are down. Turn the radio on, find the volume, and prove to me the speaker works."

Owen closed his eyes. Mark watched his son's hands. Owen fumbled for a second, then his fingers found the top knob. *Click.* The radio beeped. His thumb slid down the left side of the casing, bypassing the large PTT button, and found the smaller MONI button below it. He pressed it.

*SHHHHHHHHH!* A loud rush of static filled the kitchen. Owen rolled the volume knob down until it was a quiet hiss, then let go of the button.

"Target acquired," Owen said, opening his eyes and smirking. "Too easy."

"Perfect," Mark said. "Muscle memory. You don't need to see the radio to use it."

"My turn," Jen said, grabbing her radio and slipping on her winter coat. "Drill 3. The Clean Call. I'll go out to the cold car in the driveway."

She walked out the front door. Mark and Owen stayed in the kitchen. Mark turned the Base radio on and verified they were securely locked on Channel 3, their neighborhood simplex channel. He set the radio on the counter and stepped back.

A minute later, the squelch gate snapped open.

*"Mark, this is Jen,"* her voice came through, completely clear and steady, untainted by the howling wind outside. *"I am in the frozen minivan. Status is very cold, but secure. Ready to come back inside for coffee. Over."*

Mark grinned. She had used the exact WHO / WHERE / WHAT / NEXT format without rambling. He pressed the PTT button, paused for a full One-Mississippi count, and replied.

*"Copy that, Jen. Coffee is hot. Proceed to the kitchen. Mark clear."*

When Jen walked back inside, brushing a few stray snowflakes off her shoulder, Mark was already packing the olive-drab Go-Bag. But this time, he wasn't packing a chaotic mess.

He packed the Base radio with the extended battery attached. He packed one spare stock battery, fully charged. He packed the empty AA battery shell, pre-loaded with six Energizer Ultimate Lithiums for a grid-down backup. He tucked the laminated Quick Card and a pencil into the front zipper.

It was light. It was organized. It was in a permanent 'ready state.'

"There," Mark said, zipping the bag closed and setting it by the backdoor emergency kit. "Ten minutes. We proved we can operate in the dark, and we proved we can make a clean transmission without panicking. That's our weekly routine."

"I like this routine," Jen said, pouring her coffee. "It doesn't involve any three-hour lectures on frequency math."

"Or throat mics," Owen added, picking his earbuds back up.

Mark leaned against the counter, listening to the wind start to rattle the kitchen windows as the Alberta Clipper finally arrived. He didn't feel the urge to press buttons or test the squelch again. He didn't feel anxious. He had successfully transitioned from being a guy playing with a tactical walkie-talkie to a guy with a trusted, reliable system.

"You know," Mark said, tapping the Go-Bag with his foot. "I think we're actually ready for the snow."

## Takeaway

Skills beat features. Repetition beats panic. Do not overwhelm your family with overly complex gear, tactical accessories, or massive channel lists. Strip your Go-Bag down to the reliable essentials, ensure your radios are physically labeled, and store them *off* the chargers to protect the batteries. Stick to a simple, 10-minute weekly drill routine to keep the muscle memory sharp.

**Bridge:** Mark and Nate have both built their systems. Now, there's only one final test...

# Part 4:

## Quick Cards and Templates (Print & Laminate)

**Print It. Keep It. Use It When It Counts.**

Scan the QR code to get the printable quick cards, templates and appendix pages from this book. You'll get practical tools for setup, repeater programming, first calls, troubleshooting, and family comms planning... ready to print and keep with your radio.

https://www.morsecodepublishing.com/bootcamp

Chapter 21

# Quick Card: First Hour Checklist

**GOAL: From unopened box to your first transmission.** *If you check every box, you are successfully on the air.*

### 1. Hardware Setup

- [ ] **Attach Antenna:** Screw on the stock rubber antenna until snug. (STOP: Never press the talk button without an antenna or you risk permanent hardware damage!)
- [ ] **Seat Battery:** Slide on the battery and press firmly until it physically clicks.
- [ ] **Power On:** Turn the top knob clockwise. Set volume to roughly 50%.
- [ ] **Test Speaker:** Press the side **MONI** button to hear static. Adjust volume to comfort.

### 2. Basic Configuration

- [ ] **Set Squelch:** Press MENU -> Type 0 -> MENU -> Select 2 (this is the Goldilocks setting, quiet enough to block static, but sensitive enough to hear people) -> MENU -> EXIT.
- [ ] **Set Step Size:** Press MENU -> Type 1 -> MENU -> Select 2.5K -> MENU -> EXIT.
- [ ] **Clear Receive Tones:** Check MENU 10 and MENU 11. Both must be **OFF**.

### 3. VFO Mode (The Canvas)

- [ ] **Enter VFO Mode:** Press the orange VFO/MR button until channel numbers on the right disappear.
- [ ] **Select Top Line:** Press the blue A/B button until the tiny arrow points to the top line (A). The radio only listens to and obeys the line with the arrow!

### 4. Finding a Signal (Listen First)

- [ ] **Tune to Weather:** Type 1 6 2 . 5 5 0 (or local equivalent). Listen for the robotic voice.
- [ ] **Tune to Repeater:** Type a local Repeater Receive Frequency. Wait to hear a conversation.

**5. Programming a Repeater**

- [ ] **Clear Slot:** MENU 28 -> Select 001 -> MENU to delete old data.
- [ ] **Enter Frequency:** Type the Repeater Receive Frequency.
- [ ] **Set Offset Direction:** MENU 25 -> Select + or - -> MENU.
- [ ] **Set Offset Amount:** MENU 26 -> Type 0 0 0 6 0 0 (for 0.600) -> MENU.
- [ ] **Set Transmit Tone:** MENU 13 -> Select Tone (e.g., 100.0) -> MENU. *(Leave OFF if none required).*

**6. The Double-Save**

- [ ] **Save Receive:** MENU 27 -> Select 001 -> MENU *(Hear: "Receiving Memory").*
- [ ] **Save Transmit:** Press MENU again -> Select 001 -> MENU (Listen for the radio to say: Transmitting Memory, this is the proof your settings are saved!) -> EXIT.
- [ ] **Test It:** Press VFO/MR to enter Memory Mode. Go to Channel 001.

**7. First Transmission**

- [ ] **Check License:** Verify you hold a valid Amateur Radio License.
- [ ] **Listen:** Wait 30 seconds to ensure the channel is totally clear.
- [ ] **Talk:** Press PTT, pause for a full One-Mississippi count, say: "This is *Your Call Sign*, monitoring." Release PTT.

IF YOU ARE STUCK: Go directly to Chapters 8–12 (The Troubleshooting Tree).

Chapter 22

# Quick Card: Repeater Setup Decision Chart

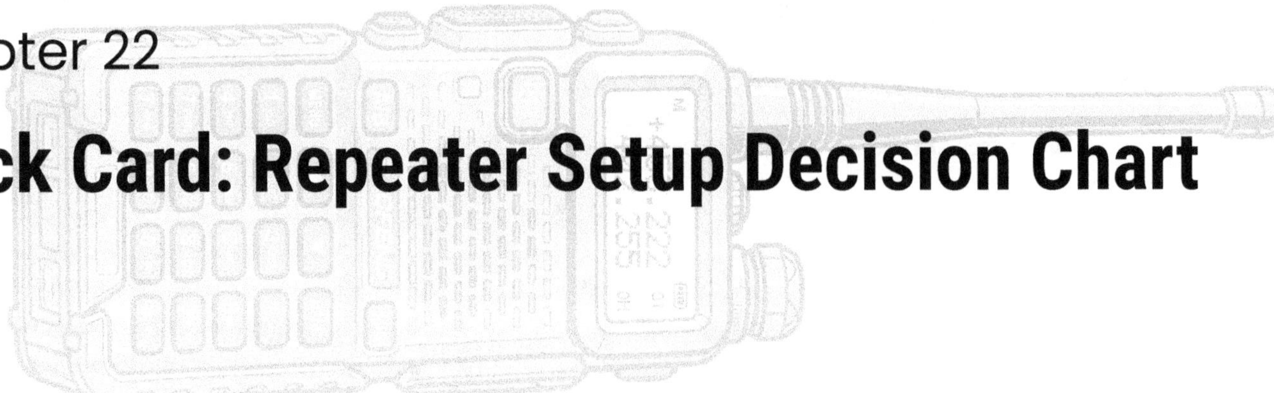

Use this flowchart when looking at RepeaterBook or a local radio club listing to know exactly which menus to change.

**Q1. Does the listing have an Offset or Shift (+ / -)?**

- **YES:** 1. Press **MENU**, type **25**, press **MENU**, select + or -, press **MENU** to confirm. 2. Press **MENU**, type **26**, press **MENU**, use the keypad to type the digits (e.g., **000600**), press **MENU** to confirm.
- **NO:** * Leave Menu 25 OFF. Leave Menu 26 at 000000. This is a simplex (direct) frequency.

**Q2. Does the listing require an "Uplink Tone" (CTCSS / PL Tone)?**

- **YES, it lists a standard number (e.g., 100.0 or 141.3):**
  - Go to **Menu 13 (T-CTCS)** and set it to that exact number.
- **YES, but it lists a "DCS" or "Digital Code" (e.g., 023 or D023N):**
  - Go to **Menu 12 (T-DCS)** and set it to that exact code.
- **NO, it says CSQ or lists no tone at all:**
  - Leave Menu 13 and Menu 12 **OFF**.

**Q3. Does the listing show a "Downlink Tone" or "Receive Tone"?**

- **DEFAULT ADVICE: IGNORE IT.** * Leave **Menu 11 (R-CTCS)** and **Menu 10 (R-DCS)** completely **OFF**.
  - *Why?* Setting a receive tone puts a lock on your own radio's speaker. Unless you are experiencing severe local interference, keeping your receive tones OFF ensures you hear everything on the channel and never accidentally lock yourself out.

**THE "SAVE IT" REMINDER**

- Did you sweep the floor? (**Menu 28: DEL-CH**). You must delete the channel slot first if it has an existing number, or the radio may not save your new offset settings correctly.
- Did you double-save? (**Menu 27: MEM-CH**). 1. Save RX: Press **MENU**, select channel, press **MENU** (Hear: *Receiving Memory*). 2. Save TX: Immediately press **MENU** again (while still on Menu 27), press **MENU** once more (Hear: *Transmitting Memory*).*

Chapter 23

# Quick Card: First On-Air Script + Etiquette

Tape this near your radio base station until muscle memory takes over.

**The 3 Starter Scripts**

**1. The "I'm Just Listening" Script** *(Use this to announce your presence without demanding a conversation.)*

"This is *Your Call Sign*, monitoring."

**2. The "Radio Check" Script** *(Use this to verify your antenna and audio are working clearly.)*

"This is *Your Call Sign*, testing a new radio setup. Is anyone available for a quick signal report?" (When they reply: "Thanks for the report, you sound great as well. *Your Call Sign* clear.")

**3. The "Quick Question" Script** *(Use this to ask for help or directions.)*

"This is *Your Call Sign*. Does anyone on frequency have a recommendation for *Your Question*?"

**The 7 Golden Rules of Etiquette**

1. **Listen First:** Never turn on a radio and immediately talk. Wait 30 to 60 seconds. Check your screen for a green RX light or signal bars, which indicate the channel is busy even if you hear silence.
2. **Press, Pause, Speak:** Press the PTT button, count "One-Mississippi" in your head, and *then* speak. This ensures the first word of your sentence isn't cut off while the repeater wakes up.
3. **Use Plain English:** Drop the CB slang. Speak naturally and clearly. Don't worry about the Q-codes (like QSL or QSY) you hear others using, plain English is the gold standard for beginners.
4. **Identify Properly:** Use your legal, government-issued call sign. By law, you must state it every 10 minutes during a long conversation and when you sign off for the last time.
5. **No "CQ" on FM Repeaters:** Do not yell "CQ, CQ, CQ." Just say your call sign and that you are monitoring.
6. **Don't Hog the Channel:** Keep transmissions under 45 seconds. Always wait for the 'repeater tail' (the click or beep after someone else stops talking) before you start your next turn.
7. **No Profanity or Music:** The airwaves are public, unencrypted, and monitored. Keep it family-friendly.

Chapter 24

# Quick Card: Family Channel Plan (Fill-in Template)

*Instructions: Fill this out with a dark marker, completely laminate with clear packing tape, and store with your radio battery.*

**FAMILY COMMS PLAN**

- **Primary Operator:** ________________________________________
- **Secondary Operator:** ________________________________________
- **Home Channel (Simplex 1):** Channel 003
- **Backup Channel (Simplex 2):** Channel 004
- **City Info Channel (Listen Only):** Channel 001

**CHECK-IN TIMES (Strict):** [ ] Top of the Hour (e.g., 2:00, 3:00) [ ] Bottom of the Hour (e.g., 2:30, 3:30) *Rule: Turn radio on 2 minutes before. Turn radio off 3 minutes after to save battery.*

**MESSAGE FORMAT: "WHO / WHERE / WHAT / NEXT"**

- **WHO:** "Dad, this is Sarah."
- **WHERE:** "I am at the high school."
- **WHAT:** "Roads are blocked. I am staying here."
- **NEXT:** "I will check back in one hour. Over."

**EMERGENCY PROTOCOL:** If in immediate, life-threatening danger, break protocol. Say: **"Emergency, Emergency, Emergency"** and give your specific location immediately.

Chapter 25

# Quick Card: Weekly Drill Tracker

*Instructions: Keep this on the fridge. Run only ONE drill per week. Keep it under 10 minutes. Muscle memory defeats panic.*

**The Drills:**

- **Drill 1 (Blindfold):** Turn radio on, set volume, use side MONI button to test speaker with eyes closed.
- **Drill 2 (Channel Switch):** Unlock keypad, switch from Channel 1 to Channel 4, re-lock keypad.
- **Drill 3 (Clean Call):** Go to different rooms. Execute a flawless "Press, Pause, Speak" using the WHO/WHERE/WHAT format.
- **Drill 4 (Lost Phone):** One person goes to a local park/store without a cell phone and checks in at the Top of the Hour via radio.

| Week | Date | Drill # | Pass / Fail | Notes (What went wrong?) |
|---|---|---|---|---|
| **Week 1** | _____ | Drill 1 | [ ] Pass [ ] Fail | ______________________________ |
| **Week 2** | _____ | Drill 2 | [ ] Pass [ ] Fail | ______________________________ |
| **Week 3** | _____ | Drill 1 | [ ] Pass [ ] Fail | ______________________________ |
| **Week 4** | _____ | Drill 2 | [ ] Pass [ ] Fail | ______________________________ |
| **Week 5** | _____ | Drill 3 | [ ] Pass [ ] Fail | ______________________________ |
| **Week 6** | _____ | Drill 3 | [ ] Pass [ ] Fail | ______________________________ |
| **Week 7** | _____ | Drill 4 | [ ] Pass [ ] Fail | ______________________________ |
| **Week 8** | _____ | Drill 4 | [ ] Pass [ ] Fail | ______________________________ |

**Note for Parents:** Start with Drill 1 and Drill 2 *only* until they are completely automatic and boring. Do not rush to Drill 4 until basic hardware mastery is achieved.

Chapter 26

# Quick Card: Fix-It Fast Cheats (Top 12 Symptoms)

*When it fails, don't panic. Check this list.*

| Symptom | Fastest Fix | Reference |
|---|---|---|
| **Keypad ignores me (beeps angrily)** | You are in Memory mode. Press **VFO/MR** to switch to manual typing. | *Ch. 8* |
| **I transmit, but friends don't hear me** | Check the **A/B** arrow. It's probably pointing to the wrong line. | *Ch. 8* |
| **Radio restarts when I press PTT** | Your battery is loose. Remove it and slam it back on until it clicks. | *Ch. 8* |
| **Tower ignores my transmission** | Your **Menu 13 (T-CTCS)** password tone is wrong or missing. | *Ch. 9* |
| **Tower ignores me (Frequency doesn't jump)** | Your Offset is OFF. Set **Menu 25** (+/-) and **Menu 26** (Amount). | *Ch. 9* |
| **My buddy is talking, but radio is silent** | You accidentally set a Receive Tone. Turn **Menu 10 & 11 OFF**. | *Ch. 10* |
| **Random blasts of loud static** | Your squelch is too low, or you are near a PC. Bump **Menu 0** to 2 or 3. | *Ch. 10* |
| **I can't type the exact frequency (rounds up)** | Your step size is too large. Change **Menu 1 (STEP)** to 2.5K. | *Ch. 10* |
| **One tower works, the other doesn't** | The broken tower uses a digital code. Turn off Menu 13, use **Menu 12**. | *Ch. 11* |
| **My saved channel disappears** | You didn't "sweep the floor." Use **Menu 28** to delete it before saving. | *Ch. 12* |
| **My saved channel works, but won't transmit** | You forgot the Double-Save. Redo **Menu 27** twice (RX and TX). | *Ch. 12* |
| **Scanner won't stop on active traffic** | Squelch is on 0 (stuck open) or 9 (stuck closed). Set it to 2. | *Ch. 13* |

# Appendices

# Appendix A: Common Terms (Plain English)

- **Baofeng:** A highly popular, inexpensive brand of handheld two-way radios manufactured in China.
- **Call Sign:** A unique alphanumeric identifier (e.g., K1ABC) issued by your government once you pass your amateur radio test. It is your official, legal name on the airwaves.
- **CTCSS/ PL Tone:** An invisible, sub-audible hum added to your voice transmission that acts as a "VIP password" to open up a repeater tower.
- **DCS:** The exact same concept as a CTCSS tone, but it uses a digital code instead of an analog hum.
- **Duplex:** The ability to transmit on one frequency and receive on a different frequency at the same time (how repeaters work).
- **Offset:** The exact mathematical difference between a repeater's listening lane and its talking lane (e.g., jumping 0.600 MHz).
- **PTT (Push-To-Talk):** The large button on the side of the radio. You hold it down to broadcast, and let it go to listen.
- **Repeater:** A large radio tower placed on a high elevation that listens for weak handheld signals and re-broadcasts them at massive power over a wide area.
- **Simplex (Direct):** Pure radio-to-radio communication without using a repeater tower.
- **Squelch:** The radio's invisible "gate" that mutes background static, only opening when a sufficiently strong human signal is detected.
- **VFO (Variable Frequency Oscillator):** The mode on your radio that allows you to manually type in raw numbers and explore the airwaves.

# Appendix B: Recommended Default Settings (Beginner-Friendly)

IF YOU GET LOST in the menus, verify these settings. This configuration creates the path of least resistance for a beginner.

- **Menu 0 (SQL):** 2 *(Filters static but catches weak signals)*
- **Menu 1 (STEP):** 2.5K *(Allows precise frequency typing)*
- **Menu 2 (TXP):** HIGH *(Maximum transmit power)*
- **Menu 3 (SAVE):** 3 *(Good battery saving ratio)*
- **Menu 7 (TDR):** OFF *(Dual Watch off prevents confusing channel jumping while learning)*
- **Menu 8 (BEEP):** ON *(Provides auditory feedback when keys are pressed)*
- **Menu 10 (R-DCS):** OFF *(Prevents locking yourself out of hearing traffic)*
- **Menu 11 (R-CTCS):** OFF *(Prevents locking yourself out of hearing traffic)*
- **Menu 14 (VOICE):** ENG *(Keeps voice prompts in English)*
- **Menu 35 (STE):** OFF *(Squelch Tail Elimination - turn off to fix weird repeater clicks)*
- **Menu 39 (ROGER):** OFF *(Nobody likes the obnoxious "roger beep" on ham bands)*

# Appendix C: Channel Naming Conventions (Examples)

When using CHIRP software to program your radio via computer, you are limited to exactly 7 alphanumeric characters for channel names. Keep them incredibly descriptive so you know what they are in a crisis.

- **CTY-RPT** (Primary City Repeater)
- **NTH-RPT** (North Backup Repeater)
- **FAM-DIR** (Family Direct / Simplex)
- **CONVOY1** (Vehicle-to-Vehicle Travel Channel)
- **WX-CH1** (Weather Broadcast 1)
- **FIRE-DS** (Local Fire Dispatch - Receive Only)

# Appendix D: Getting Local Repeaters (Where to Look)

Do not guess frequencies. The easiest way to find accurate, up-to-date repeaters in your specific neighborhood is:

1. **RepeaterBook:** Download the free "RepeaterBook" app on iOS or Android, or visit www.repeaterbook.com. It uses your GPS to show you exactly how many miles away the nearest tower is.

2. **Local Clubs:** Google "[Your County] Amateur Radio Club." They almost always list their primary repeater on the homepage.

**What to Copy Down on Your Notepad:**

1. The **Downlink / Receive** Frequency (e.g., 146.700).

2. The **Offset** Direction (+ or -).

3. The **Uplink Tone** (e.g., 100.0). Ignore the Downlink tone if they list one.

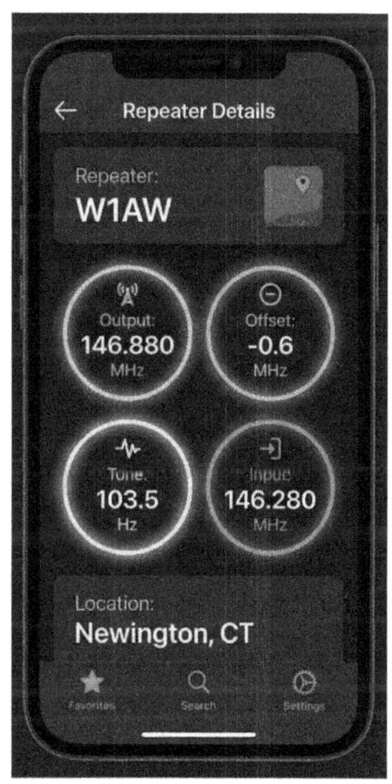

# Appendix E: Legal Basics (What You Can / Can't Do)

- **Listening is Free:** In the US (and most countries), it is 100% legal to listen to almost any unencrypted frequency without a license.
- **Transmitting is Governed:** To press the PTT button on Amateur Radio (Ham) bands, you **must** hold a valid license from your government (like the FCC).
- **The Exception:** Fines are steep for illegal transmitting. The *only* exception generally recognized is an immediate, genuine threat to life and limb where no other method of communication is available. A flat tire does not count. A life-threatening medical emergency during a total grid collapse does.
- **Get Licensed:** The entry-level "Technician" license in the US is a 35-question multiple-choice test. It requires zero morse code and costs about $15 to take. Spend the time to study, take the exam, and unlock the full legal power of your Baofeng.

# Appendix F: Emergency Reference (Stay Calm)

IF THE GRID GOES down right now:

1. **Do not transmit immediately.** Turn the radio on, tune to your City Repeater (Channel 1), and simply listen.
2. **Gather Intel.** Local emergency nets will establish a "control station." Write down what they are saying (road closures, shelter locations).
3. **Switch to Simplex.** If you need to find your family, switch to your Family Direct channel (Channel 3).
4. **Use WHO/WHERE/WHAT.** Keep transmissions under 10 seconds.
5. **Conserve Power.** Turn the radio off when not actively listening for scheduled check-ins. Keep your radio warm (inside your jacket) if you are in a freezing environment to preserve battery chemistry.

# A Request: Keep the Hobby Growing

If this book helped you finally understand your Baofeng radio and gave you the confidence to make your first transmission, please consider leaving a review where you purchased it.

More importantly, help the next person. When you hear a nervous, shaky voice on your local repeater saying, *"This is my first time on the air,"* be the person who comes back, welcomes them, and gives them a fantastic signal report.

*Stay prepared. Stay on the air.*

# Narrative Interlude: Two Weather Watches

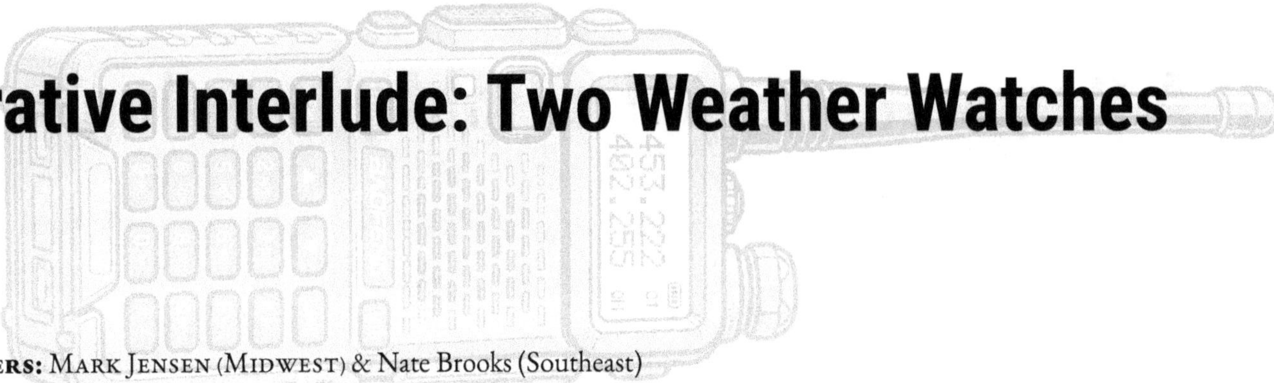


**Characters:** Mark Jensen (Midwest) & Nate Brooks (Southeast)

**Skills reinforced:**

- Quick cards under stress
- Triage mindset
- Family plan execution

**Where to learn the steps:** Parts 2, 3, & 4

At 6:14 PM in the Minnesota suburbs, the neighborhood transformers began to fail. It sounded like a string of distant, muffled fireworks popping off under a thick blanket. Mark Jensen stood in his living room, staring out the front window as the heavy, blinding curtain of an Alberta Clipper furiously buried his driveway in white.

With a final, loud metallic *pop* from the pole at the end of the street, the regional power grid surrendered. The comforting, continuous hum of the furnace spun down to a halt, and the house was instantly plunged into absolute, freezing darkness. The silence that followed was heavy and immediate.

"Wi-Fi is down!" Owen shouted from the basement, his voice laced with the sudden, primal panic of a disconnected teenager whose digital world had just vanished. "Dad, the router is completely dead!"

"Everyone stay put! Nobody move until we get lights!" Mark called out.

He didn't scramble in the dark. He didn't frantically tear through the kitchen junk drawers looking for loose AA batteries or half-dead flashlights. He walked with measured, practiced steps to the hallway closet. He reached in, unzipped the front pouch of his neatly staged olive-drab Go-Bag, and pulled out his Baofeng radio and a small, high-lumen LED flashlight. He clicked the light on, instantly sweeping away the darkness.

Twelve hundred miles to the south, the situation was much louder and infinitely wetter.

Nate Brooks stood in his garage in South Carolina, the air thick with barometric pressure and the smell of wet pine needles. He was listening to the outer bands of a Category 3 hurricane actively punishing the reinforced aluminum of his garage door. The wind didn't howl; it roared like a freight train idling right in his driveway. The power had been out for three agonizing hours. The cell networks, overwhelmed by a million people trying to stream weather updates, had overloaded and died thirty minutes ago. His phone screen simply displayed the useless, terrifying phrase: *SOS Only*.

Kara had gone down the street to check on an elderly neighbor right before the cell towers failed, intending to be back in ten minutes. That was half an hour ago.

Nate stood at his workbench, his face illuminated by the harsh, white glare of a single battery-powered lantern. Sitting perfectly centered next to the lantern was his Baofeng UV-5R. The bright yellow, laminated Quick Card was tied securely to its pouch, an anchor of procedure in a chaotic environment.

Back in Minnesota, Mark clicked the top knob of his radio.

*"BEEP. CHANNEL MODE."*

"Dad," Owen said, appearing at the top of the basement stairs, squinting against the flashlight beam with his own radio clutched in his hand. "I think I broke mine in the dark. I bumped the keypad when the lights went out, and now it's just hissing at me. I can't get it to stop."

Mark felt a brief, phantom spike of his old beginner's anxiety, the deep-seated fear of the beeping brick, the assumption that cheap electronics had failed exactly when he needed them most. But then his training kicked in. He remembered the triage flow.

"Bring it here," Mark said, aiming his flashlight at the green LCD screen of Owen's radio. He ran the 60-second mental checklist, refusing to panic. *Mode?* He saw the tiny numbers on the right. Memory mode. *Arrow?* It was pointing to the top line. *Squelch?* Mark looked closely at the screen. The tiny "SQL" icon wasn't showing, meaning the radio wasn't receiving a signal. Then, he looked at the side of the chassis. He noticed the small rubber button below the PTT was physically wedged against the edge of Owen's nervously gripping thumb.

"You're holding down the Monitor button, bud," Mark chuckled. He felt a wave of profound relief, mostly because he knew he would have assumed the radio was fried just two weeks ago. He gently pulled Owen's thumb off the side of the radio. The rushing waterfall of static instantly stopped, plunging the hallway back into quiet. "The invisible gate was open. You just bypassed the squelch. You're fine. The radio works perfectly."

Mark checked his own radio. He was locked onto Channel 5, the National Weather Service frequency. He listened for thirty seconds to confirm the blizzard warning hadn't escalated, then confidently clicked the Up Arrow down to Channel 3, their designated Family Simplex channel.

"Okay, Owen," Mark said, switching seamlessly back into Mission Control mode. "Take your radio. Go back down and check the basement egress windows for snow drafts or leaks. I'll stay up here and check the front doors. Let's do a clean call in two minutes."

In South Carolina, Nate Brooks checked his tactical watch. It was 14:58. Two minutes to the top of the hour.

He didn't touch the radio. He didn't pace the garage floor. He simply stood still and waited. The urge to break protocol and call Kara immediately was a physical weight in his chest, but he resisted it. Radio discipline was the line between control and chaos. The plan was built to eliminate the anxiety of the unknown, and he trusted the procedure above his own nerves. He knew that Kara had her radio in her jacket pocket, perfectly tuned to Channel 3, and he knew she wouldn't waste her battery trying to call him outside of the designated window.

At exactly 15:00, Nate reached out, his hand steady, and clicked the top knob on. He verified the green screen read Channel 3. He set it back down on the bench.

At 15:01, the green LED indicator light on the top of the radio flashed. A split-second later, the squelch gate snapped open. It was a loud, aggressive burst of static that cut cleanly through the muffled roar of the hurricane outside.

*"Nate, this is Kara,"* her voice came through the small speaker. It was slightly scratchy, battling the dense, wet foliage of the storm, but perfectly legible. *"I am at Mrs. Gable's house. Everyone is secure. Waiting out this band of rain before walking back. Over."*

Nate picked up the radio. He pressed the PTT button, counted a deliberate *One-Mississippi* in his head to let the hardware wake up and open the channel, and replied with a steady, flat cadence.

*"Copy that, Kara. Stay put until the rain slacks. I am monitoring Channel 3. KN4XYZ clear."*

He set the radio down. The transmission had taken eight seconds. It was entirely devoid of panic, perfectly legal, and completely off the grid. The feeling of absolute, quiet confidence that washed over Nate in that dim garage was profound. He had taken a thirty-dollar piece of plastic and turned it into a reliable lifeline. The system wasn't just theoretical anymore. It worked.

In Minnesota, Mark's radio chirped cheerfully from the kitchen counter, cutting through the eerie silence of the dead house.

*"Dad, this is Owen,"* the voice said, sounding surprisingly professional and composed for a sixteen-year-old in a blackout. *"I am in the basement. Windows are dry, no snow getting in. It's really cold down here, though. Over."*

Mark grinned in the dark, incredibly proud of his son for sticking to the script. He picked up his radio, pressed the PTT button, and paused.

*"Copy that, Owen. Come back upstairs and grab a heavy blanket from the closet. Good work. Mark clear."*

Jen walked into the kitchen, wrapped in a heavy wool sweater, holding a flickering, battery-operated candle. She looked at Mark standing tall in the dark, then at the radio resting next to the printed, laminated Comms Plan.

"Are we officially surviving the snowpocalypse?" she asked, raising an eyebrow with a fond smile.

"We have continuity of operations, Jen," Mark said, unable to hide his grin. "The NASA control room is fully functional. We have eyes on the basement."

"It's a thirty-dollar walkie-talkie, Mark. You are standing in a coat closet."

"It's a verified, multi-tiered communication node," Mark corrected gently, locking the keypad with a precise two-second press of the # key. "And more importantly, it works."

Hundreds of miles apart, battling completely different elements on opposite sides of the country, both men sat in the dark and listened to the ambient noise of their respective storms. The fear of isolation, that creeping, helpless dread that usually accompanies a sudden loss of the cellular grid and the internet, simply wasn't there.

They weren't electrical engineers. They didn't have to memorize complex physics, build massive backyard towers, or understand the intricate internal wiring of their digital devices. They had just followed the steps, built their simple Quick Cards, and practiced their lines before the crisis hit.

They had taken a confusing, beeping piece of plastic and turned it into a trusted lifeline.

They were ready.

## Takeaway

The ultimate goal of amateur radio and emergency preparedness is peace of mind. You don't need a hundred programmed channels, a trunk full of expensive tactical gear, or an advanced degree to protect your family. You need a charged battery, a simple, laminated one-page plan, and the discipline to practice it before the lights go out.

You can do this. Turn that beeping brick into a lifeline. Stay prepared, and stay on the air.

# Build Your Library

## The Operator's Upgrade Path

You have mastered the Baofeng. You have the hardware, the channel plan, and the mindset. To truly secure your connection to the outside world, you need to expand your legal privileges, your range, and your hardware options.

Here are three guides from **Morse Code Publishing** designed to take you from "Guy with a Radio" to "Communications Commander."

### 1. GET LEGAL, GET LOUD

#### Ham Radio Technician Class Study Guide

You know how to use the radio now. But do you have the credentials to use it openly?

Operating "under the radar" is fine for the apocalypse, but it prevents you from practicing on Tuesday afternoons. To truly test your range, join local repeater nets, and build a network of skilled friends, you need your Technician License.

Don't let the exam scare you. You don't need a degree in electrical engineering. This guide strips away the confusing jargon and focuses on the no-nonsense facts you need to pass.

**The Hook:** Stop worrying about the FCC and start transmitting with authority.

**Best For:** The operator who wants to join the local ARES/RACES emergency groups.

## 2. WHEN LOCAL ISN'T ENOUGH

### Ham Radio General Class License Study Guide

Your Baofeng is a "Line of Sight" tool. It dominates the neighborhood. But what happens if a hurricane wipes out the power grid for the entire state? What if the local repeaters are dead?

You need to reach the outside world.

Upgrading to your **General Class License** unlocks the High Frequency (HF) bands. This is the "Big League" of radio—bouncing signals off the ionosphere to talk to Texas, Maine, or Europe without satellites or the internet. This guide takes the intimidation out of the General exam, breaking down the math and theory into plain English.

**The Hook:** Break the line-of-sight barrier. Secure your link to the outside world.

**Best For:** The prepared citizen who needs regional and national reach.

## 3. THE "NO-TEST" FAMILY SOLUTION

### GMRS Radios: The Visual Command System

Let's be honest: You might be willing to study for a radio exam, but your spouse and teenagers probably aren't.

**GMRS (General Mobile Radio Service)** is the answer. One $35 license covers your **entire immediate family**—no test required.

This manual is the perfect companion to your Baofeng setup. It focuses on high-power mobile units for your vehicles and simplified protocols for the family. It teaches you how to build a 50-watt mobile "base station" in your truck and how to integrate GMRS into your family's safety plan without the friction of exams.

**The Hook:** High-power comms for the whole tribe, zero studying required.

**Best For:** The family fleet, road trips, and overlanding.

www.ingramcontent.com/pod-product-compliance
Lightning Source LLC
Chambersburg PA
CBHW061356240926
51846CB00007B/1426
* 9 7 8 1 9 6 3 1 4 2 5 0 1 *